REPAIR & RENOVATE

masonry&plastering

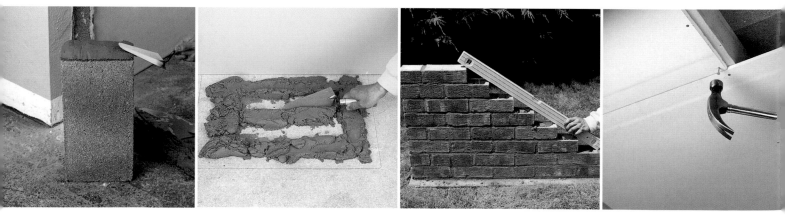

REPAIR & RENOVATE

masonry&plastering

Mike Lawrence

MURDOCH
B O O K S

masonry & plastering
contents

See pages 112–13 for how to patch up damaged plaster.

See pages 106–7 for how to attach a ceiling rose.

introduction

Home maintenance, restoration and improvement are major leisure-time activities for many home owners, who are becoming more and more adept at tackling jobs traditionally left to the professionals. People are prepared to learn skills in areas such as building and masonry projects, that were once seen as too difficult to master successfully. There is just as much pleasure in mastering bricklaying and plastering as there is in painting a ceiling or putting up a shelf – all it takes is practice.

doing your homework

Before embarking on any masonry or plastering project, it pays to take some time examining the structure of your house, and so the book begins with a look at the anatomy of masonry structures. This chapter identifies the main types of construction – old and modern, solid and cavity, timber and block – for the external and internal walls of a house, as well as the construction of ceilings, and also demonstrates the options available for garden walls, patios and paths. With this information you will be able to identify the masonry structures that already exist in your home so

that you can better understand how your house was built. This in turn will make it easier to plan and execute any construction or alteration work you want to carry out on it. If you are thinking about improving your garden by building walls or laying a patio or driveway, you will also need to know which materials to select and what construction methods to use for garden walls and paved surfaces. Armed with this invaluable background knowledge, you will then be better equipped to tackle the task ahead and can move onto the planning stage.

BELOW *This brick garden wall with stone block coping acts as a neat retaining wall for an elevated flowerbed.*

ABOVE *Where the walls in a room are decorated with a simple covering of paint, it is particularly important to ensure you achieve a perfect finish when plastering.*

planning the job

The subject matter of Chapter 2 covers the several stages of planning required. The first is to decide exactly what you want to achieve. You can get plenty of ideas for indoor projects by looking at home interest magazines and interiors books, while gardening magazines and landscape gardening books will provide inspiration for redeveloping outdoors. The other part of the planning equation involves making sure that you have (or can borrow or hire) the right tools and equipment, including ladders, steps and work platforms, so that you are fully equipped to carry out any job effectively and with the greatest ease – there is no sense making things more difficult for yourself by using the wrong tools. You also need to choose your materials and estimate quantities, then decide where to buy them and how you intend to store them. Last but by no means least, you need to plan the order of work and check whether any official consent is required – either planning permission or Building Regulations approval – before you can proceed.

making changes

Chapters 3, 4 and 5 look in detail at some of the most popular masonry projects. Inside the house, there is a wide range of improvements you may consider tackling. You might want to divide a large room by building a block partition wall, or make two rooms into one by removing an internal masonry wall. Creating a new door opening will enable you to improve traffic routes and make better use of floor space, while inserting a new window can dramatically improve the way the room receives daylight. Conversely, blocking up an unwanted window or door opening can help to give you extra wall space for furniture. Lastly, you might want to recommission an old, unused fireplace, block off an unwanted one or add a new masonry fire surround to give your room an imposing focal point.

Outdoors is an excellent place to hone your masonry skills without having to worry about making a mess and disrupting the household. There is a wealth of projects you can take on, from building walls in brick, stone or man-made blocks to creating a brick arch or making a flight of steps if you have a sloping site or a raised patio. When it comes to creating access or seating areas, you can consider laying paving slabs, block pavers or crazy paving and adding areas of gravel or cobbles. Or you could simply lay inexpensive and practical concrete, whether for a driveway or path or as a base for a garden shed.

plastering

Without a doubt, plastering – the subject of chapter 6 – is one of the most difficult skills to master. As a result, a good plasterer can command the highest rate of pay of any professional – as good a reason as any to master the plasterer's craft and save yourself a great deal of money. The job involves plastering masonry, working with and plastering plasterboard on walls and ceilings, and dry lining walls to create smooth new surfaces for decorating where the existing walls are in poor condition. There are a number of related jobs that are much easier to tackle yourself, including creating an archway in an existing opening, fitting ceiling coving and adding ornamental plaster features such as ceiling centres and panel mouldings. All these add detail and distinction to any room.

BELOW *Various masonry materials are suitable for laying as a patio. Here several sizes of textured slabs have been laid on a sand base to create a stylish seating area in the garden.*

repair & renovation

Perhaps the least enjoyable masonry and plastering jobs are those that involve fixing faults and curing problems. However, carrying out repairs as soon as possible after the problem appears is a vital part of maintaining both the appearance and structural integrity of your home. The final chapter looks in detail at the most common repairs to masonry and plastering you may need to make. Indoors, you are likely to be called upon to patch plaster, fill cracks in walls and ceilings, repair damage to plasterboard surfaces and even to make good a crumbling concrete floor. Outdoors, regular maintenance jobs include repointing brickwork, replacing damaged bricks and rendering, patching cracked concrete and levelling paving slabs and blocks that have subsided as time goes by. If you are unlucky you may need to do battle with rising damp.

Whatever jobs and projects you decide to take on, you will improve your skills and develop talents you did not know you had. With luck, you will come to realize that masonry and plastering hold no fears after all. The book closes with a glossary of the technical terms used in the text, and a list of contacts if you need technical or professional assistance before or during your project.

The layout of this book has been designed to give project instruction in as comprehensive yet straightforward a manner as possible. The illustration below provides a guideline to the different elements incorporated in the page design. Colour photographs and diagrams combined with explanatory text, laid out in a clear, step-by-step order, provide easy-to-follow instructions. Each project is prefaced by a blue box containing a list of tools so that you will know in advance the range of equipment required for the job. Other boxes of additional text accompany each project, and are aimed at drawing your attention to particular issues. Pink safety boxes alert the reader to issues of safety and detail any precautions that may need to be taken. They also indicate where a particular job must be carried out by a tradesperson. Green tip boxes offer professional hints for the best way to go about a particular task involved in the project. Boxes with an orange border describe alternative options and techniques, relevant to the project in hand but not demonstrated on the page.

difficulty rating

The following symbols are designed to give an indication of difficulty level relating to particular tasks and projects in this book. Clearly what are simple jobs to one person may be difficult to another, and vice versa. These guidelines are primarily based on the ability of an individual in relation to the experience and degree of technical ability required.

Straightforward and requires limited technical skills

Straightforward but requires a reasonable skill level

Technically quite difficult, and could involve a number of skills

High skill level required and involves a number of techniques

a list of tools is provided at the beginning of each job

option boxes offer additional instructions and techniques for the project in hand

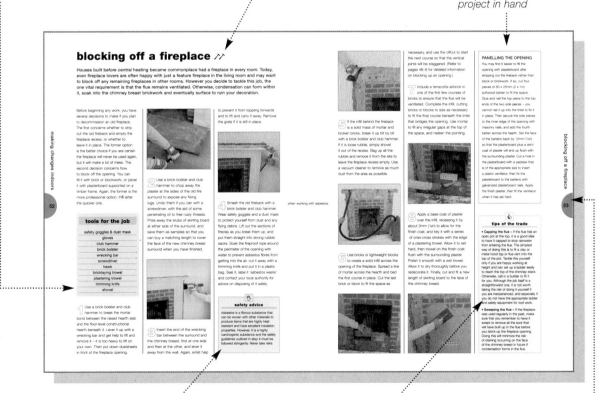

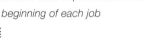

blocking off a fireplace

Houses built before central heating became commonplace had a fireplace in every room. Today, even fireplace lovers are often happy with just a feature fireplace in the living room and may want to block off any remaining fireplaces in other rooms. However you decide to tackle this job, the one vital requirement is that the flue remains ventilated. Otherwise, condensation can form in it, soak into the chimney breast brickwork and eventually surface to ruin your decoration.

tools for the job

safety goggles & dust mask
gloves
club hammer
brick bolster
wrecking bar
screwdriver
hawk
bricklaying trowel
plastering trowel
trimming knife
shovel

safety boxes, pink for emphasis, draw attention to safety considerations

tip boxes provide helpful hints developed from professional experience or highlight areas where more traditional methods can be used

colour-coordinated tabs help you find your place quickly when moving between chapters

anatomy of masonry structures

Before you can contemplate carrying out any repair or renovation work on the various masonry structures that make up your home and its surroundings, you need a basic understanding of how they are put together. House wall construction has evolved over the years, from solid masonry to cavity walls and timber framing. In addition, internal partitions and ceilings in older properties will be very different from those in recently built houses. Outside the house, there are more ways to build a garden wall than you might imagine, and now that home improvement has become a major leisure-time activity, new materials have come into vogue for paved areas such as drives, paths and patios. This chapter describes the most common construction techniques used for all these masonry elements, to enable you to recognize what you have and what you want to create in your home.

The brickwork for both this garden wall and wall of the house share the same stretcher bond pattern.

external house walls

A close look at the outside walls of your house may reveal everything or nothing about how they were constructed. Exposed stone or brickwork gives plenty of clues, but if building materials are concealed behind cement rendering, timber cladding or tile hanging, you have to do a little more detective work to discover what lies beneath the surface. The descriptions here, and an estimate of your house's age, will in most cases enable you to work out how your walls are constructed.

solid walls

Any house built before about 1920 is almost certain to have solid masonry external walls. These may be built with random or regular stone blocks, or with bricks. The thickness of brick walls gives a clue to their construction. Most two-storey houses have walls one brick (230mm/9in) thick, and three-storey houses may have ground-floor walls one-and-a-half bricks (350mm/13¾in) thick. If the bricks are exposed on the outside, the way they are arranged side-on and end-on in successive layers (courses) will also give a clue to the way the wall was built.

cavity walls

Cavity walls were introduced from around 1920 to improve the weather resistance and thermal insulation of external walls. They consist of two layers of masonry called leaves, usually half a brick (115mm/4½in) thick each, with a gap between them. The outer leaf is generally made of bricks, while the inner leaf may be of brick or block construction. In timber-framed cavity walls, the outer leaf is of masonry while the inner leaf is formed from a series of prefabricated timber-framed panels.

The gap between the leaves not only traps air and improves the insulating properties of the wall, but also channels any water that penetrates the outer wall down its inner face to ground level, preventing the rain penetration that is a problem with solid walls, especially in exposed areas. Standard cavity walls have a 50mm (2in) wide gap, giving an overall wall thickness of about 280mm (11in). If the bricks are exposed on the outside, only brick sides (called stretchers) and the occasional cut brick will be visible.

INTERNAL FINISHES

• **Solid brick** – Coated with cement rendering (mortar) and then given a finish coat of lime plaster. The coating may be as much as 25mm (1in) thick.

• **All-brick cavity** – Probably plastered, using an undercoat and topcoat of quick-setting gypsum-based plaster. The coating will be no more than 13–16mm (½–⅝in) thick.

• **Brick-and-block cavity** – May be rendered and plastered as for solid walls, or have two layers of lime plaster. Again, the coating may be as much as 25mm (1in) thick.

• **Timber-framed cavity** – Lined with plasterboard, which sounds hollow when tapped, a sure indicator of a timber-framed wall. The plasterboard may then be plastered or dry lined.

solid brick wall

Bricks are laid in one of several different patterns (called bonds) that interlock to give the wall its strength (see page 16). The most common are called English and Flemish bonds (the latter is illustrated here). The bricks are laid overlapping each other so that no vertical joints coincide. The long side of a brick is called a stretcher, and the short end a header. Both are visible in regular patterns across the outside face of the wall.

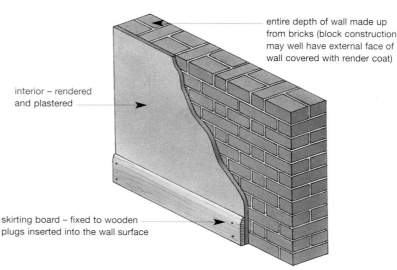

interior – rendered and plastered

entire depth of wall made up from bricks (block construction may well have external face of wall covered with render coat)

skirting board – fixed to wooden plugs inserted into the wall surface

13

all-brick cavity wall

The two brick leaves are separated by a gap – usually 50mm (2in) wide – and are tied together by metal (usually lead) ties bedded in the mortar between the bricks. Each leaf consists of bricks laid end to end so that only their long sides (stretchers) are visible on the outside – a bond known as stretcher bond, and an excellent clue to the structure of the wall. The roof is supported by the inner leaf of the wall.

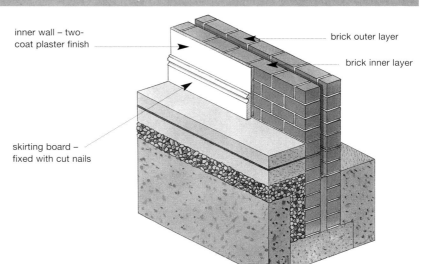

inner wall – two-coat plaster finish

brick outer layer

brick inner layer

skirting board – fixed with cut nails

brick-and-block cavity wall

The outer leaf is brick or reconstituted stone in brick shapes, laid in stretcher bond. The inner leaf is built of larger blocks – concrete or breeze blocks in older homes, and lightweight insulating blocks (Thermalite or similar) in newer ones to improve thermal insulation. The two are tied together with twisted steel, galvanized wire or plastic ties. The cavity may be filled with insulation, either during construction with slabs of glass-fibre insulation blanket, or afterwards with an injected insulation material.

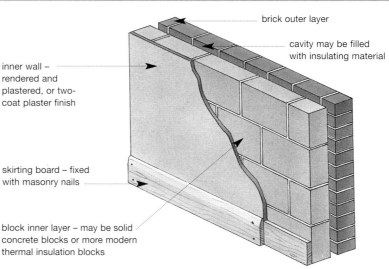

brick outer layer

cavity may be filled with insulating material

inner wall – rendered and plastered, or two-coat plaster finish

skirting board – fixed with masonry nails

block inner layer – may be solid concrete blocks or more modern thermal insulation blocks

timber-framed cavity wall

The outer leaf is the same as for a brick cavity wall. The inner leaf is built of timber-framed panels faced on the cavity side with exterior-quality (water-resistant) plywood. The spaces within the frame are filled with slabs of glass-fibre insulation, which are covered on the inner side with a polythene vapour barrier to prevent moist air from inside the house causing condensation in the insulation. Galvanized steel ties are screwed to the cavity side of the panels and bonded into the outer leaf at intervals to tie the two together.

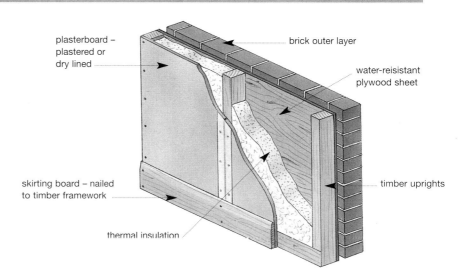

plasterboard – plastered or dry lined

brick outer layer

water-reisistant plywood sheet

skirting board – nailed to timber framework

timber uprights

thermal insulation

internal house walls & ceilings

Some internal walls are simple room dividers, while others are loadbearing. Downstairs, a loadbearing wall carries the weight of upstairs floor joists and may also support an upstairs wall directly above it. This upstairs wall may in turn support part of the roof structure. Depending on the age and construction of the house, these walls may be of masonry or of timber-frame construction. Ceilings, by contrast, show less variety.

solid walls

In houses with masonry external walls, internal ground-floor walls are usually also of masonry. Brick is commonplace in houses built before the 1930s, when concrete blocks began to take their place – concrete is now used universally for all solid loadbearing internal walls. Upstairs, loadbearing walls are upward continuations of similar ground-floor walls, and will also be of masonry.

hollow walls

Timber-framed walls are widely used upstairs where their positions do not coincide with the location of ground-floor walls. Instead, their weight is carried by the upstairs floor joists. Such walls are not loadbearing, but other timber-framed walls in the house may be. Only a close examination will determine whether the wall is supporting floor joists or roof struts.

timber-framed walls

In houses built before the mid-1920s, lath-and-plaster walls are the norm. Their frame consists of vertical members called studs fixed between a horizontal head and sole plate. The spacing of the studs is usually around 610mm (24in). Long, thin strips of softwood, called laths, are nailed horizontally across the studs with a narrow gap between them, and a first coat of lime plaster is applied. This is squeezed between the laths to form a key for subsequent coats of plaster.

Since the introduction of plasterboard in the 1920s, timber-framed walls have become the most common type of internal wall. The framing is similar to a lath-and-plaster wall, with the addition of horizontal braces called noggings between the studs to prevent them from twisting. Plasterboard is sold in standard-sized sheets, so the stud spacing is critical to allow board joints to be centred over them. They are normally fixed at 400mm (15¾in) centres, occasionally at 600mm (24¾in). The board surface may be given a thin skim coat of plaster, or have the joints taped and filled – a finish known as dry lining.

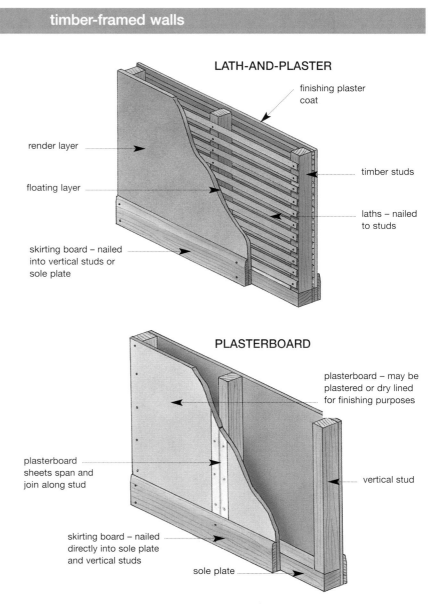

LATH-AND-PLASTER

finishing plaster coat

render layer

floating layer

timber studs

laths – nailed to studs

skirting board – nailed into vertical studs or sole plate

PLASTERBOARD

plasterboard – may be plastered or dry lined for finishing purposes

plasterboard sheets span and join along stud

vertical stud

skirting board – nailed directly into sole plate and vertical studs

sole plate

solid brick or block wall

Internal brick walls are usually one brick (102mm/4in) thick, with bricks laid in a stretcher bond arrangement. The wall surfaces will be rendered and plastered on both sides, giving an overall thickness of 140mm (5½in), though buildings of three storeys or more may have thicker ground-floor walls. Concrete block walls are one block thick – generally 100mm (4in) – with a 13mm (½in) coat of plaster on each face, giving an overall wall thickness of 125mm (5in).

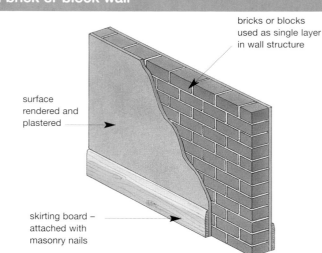

bricks or blocks used as single layer in wall structure

surface rendered and plastered

skirting board – attached with masonry nails

ceilings

In houses with timber floor joists, the ceiling structure is fixed to the undersides of the joists and the floor of the room above is laid on top of them. Lath-and-plaster ceilings are the norm in houses built before the 1920s and 1930s, since when plasterboard ceilings have become universal. Lath-and-plaster ceilings are constructed in the same way as lath-and-plaster walls, and have a three-coat lime plaster coating, while plasterboard ceilings may be skim-coated with finish plaster or dry lined as for plasterboard walls. Joist spacings may vary in older homes, but are almost always at 400mm (15¾in) centres in modern ones. Timber ceilings of tongue and groove boarding are occasionally found in Victorian houses.

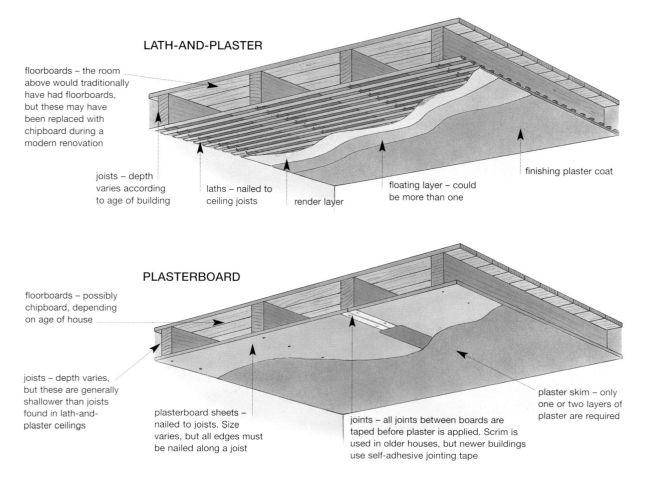

LATH-AND-PLASTER

floorboards – the room above would traditionally have had floorboards, but these may have been replaced with chipboard during a modern renovation

joists – depth varies according to age of building

laths – nailed to ceiling joists

render layer

floating layer – could be more than one

finishing plaster coat

PLASTERBOARD

floorboards – possibly chipboard, depending on age of house

joists – depth varies, but these are generally shallower than joists found in lath-and-plaster ceilings

plasterboard sheets – nailed to joists. Size varies, but all edges must be nailed along a joist

joints – all joints between boards are taped before plaster is applied. Scrim is used in older houses, but newer buildings use self-adhesive jointing tape

plaster skim – only one or two layers of plaster are required

garden walls

Garden walls can mark property boundaries, act as dividers between different levels of the garden on sloping sites, or form self-contained planters and other garden features. Their method of construction varies according to whether their purpose is structural or merely decorative, and whether you are building in brick or blocks. If you are building a garden wall from scratch, its foundations below ground level are as important as its structure above.

foundations

Even the lowest garden wall needs proper in-ground support. In most cases this means excavating topsoil and subsoil, and laying a concrete strip foundation using a mix of 1 part cement to 5 parts combined aggregate (mixed sharp concreting sand and 20mm/³/₄in gravel). The strip should be twice as wide as the wall for masonry up to 750mm (29½in) high, and three times the width for higher walls. It should be at least 150mm (6in) thick on all soil types except clay, in which case it should be 200mm (8in) thick because clay is prone to subsidence. Its top surface should be about 230mm (9in) below ground level, so that three courses of bricks or one of blocks can be placed below ground level and planting or turf can be laid next to the wall.

brick walls

The simplest brick garden walls are built as a single layer of brickwork, one brick (102mm/4in) thick. The bricks are laid in stretcher bond, with each brick overlapping those in the course below by half its length. The maximum safe height for this type of construction is 450mm (17³/₄in) – six courses of bricks – unless it is supported by piers one brick square at the ends and at 3m (10ft) intervals in between, when a further three courses can be added. A taller wall needs brickwork twice as thick – the length of a brick (215mm/8½in) instead of the width of one. You can build a wall like this to a height of 1.35m (4ft) without piers, and to 1.8m (6ft) with end and intermediate piers one-and-a-half bricks (about 330mm/13in) square. Building a thicker wall means having to adopt a different way of arranging the bricks (see box).

TYPES OF BOND

Three different bonding arrangements are commonly used for garden walls.

- **English bond –** The first course has two rows of parallel stretchers, the second course is formed by headers (bricks laid end-on to the face of the wall). Alternate stretcher and header courses build up the wall. In header courses, a brick split lengthways (queen closer) is used at corners.

- **Flemish bond –** Each course has two parallel stretchers followed by one header. The headers are centred over the pair of stretchers in the course below. Queen closers are used at the ends and corners in alternate courses.

- **English garden wall bond –** The first three, four or five courses are laid as parallel stretchers, followed by a single course of headers. This type of bond is similar to English bond, but it is not as strong because there are not so many courses of headers.

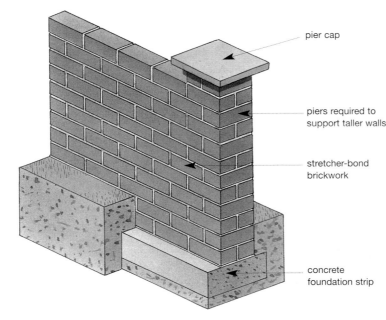

pier cap

piers required to support taller walls

stretcher-bond brickwork

concrete foundation strip

Two types of manufactured blocks are commonly used for garden walls. The first is reconstituted stone. These blocks have one face and one end moulded to resemble natural stone, while the other faces are smooth so the blocks can be laid just like bricks in level courses. Some ranges include blocks of different lengths and heights, creating the appearance of random stonework when laid. The same height restrictions apply as for brickwork, with the provision of piers being necessary for high walls.

The second type is the pierced screen wall block. This is a square block pierced with a variety of simple designs. They are commonly 300mm (11¾in) square and 90mm (3½in) thick, and are designed to be laid in stack bond – in columns and rows with no overlap between the blocks, which obviously cannot be cut to size. The resulting wall is comparatively weak unless piers are built at 3m (10ft) intervals and metal mesh reinforcement is used in the horizontal mortar joints. Piers can be of brick or solid block, or can be constructed using special matching end, corner and intermediate pier blocks that are 200mm (8in) high and have hollow centres so that they can be erected around reinforcing rods set in the foundation concrete. The maximum wall height is 600mm/23½in (two blocks) without reinforcement, and 1.8m (6ft) with it.

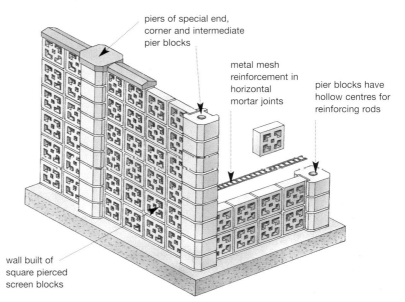

piers of special end, corner and intermediate pier blocks

metal mesh reinforcement in horizontal mortar joints

pier blocks have hollow centres for reinforcing rods

wall built of square pierced screen blocks

Walls used to terrace a sloping garden have to be strong enough to hold back the weight of earth. A low brick wall up to 600mm (23½in) high can be built in 215mm (8½in) thick brickwork, without reinforcement, in one of the bonds described opposite. For a wall up to 1.2m (4ft) high, the wall is built as two layers in stretcher bond, with reinforcing rods set in the foundations and sandwiched between the two walls. For extra strength, the walls are bonded together with cavity wall ties, and the cavity is filled with fine concrete. Walls higher than 1.2m (4ft) must be built by professionals to ensure that they are strong enough to withstand collapse. For this reason, it is better to terrace a sloping site with several low walls forming a series of shallow tiers, rather than to have a single high retaining wall.

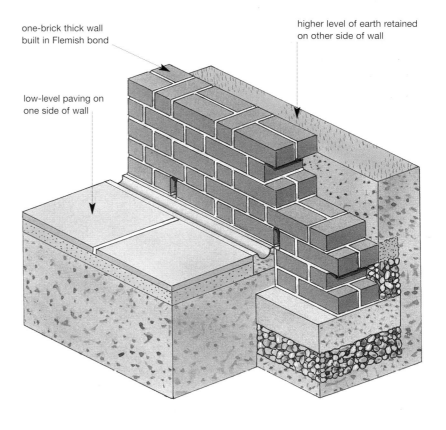

one-brick thick wall built in Flemish bond

higher level of earth retained on other side of wall

low-level paving on one side of wall

patios, paths & drives

Apart from walls, garden landscaping also requires hard surface areas for traffic and to take garden furniture. Patios, paths and drives can be surfaced in a variety of materials, including individual slabs or blocks laid on a sand or mortar bed, and areas of concrete, cobbles or gravel. The key requirements of any hard surfaces outside the house are that they should provide a stable, firm surface that drains freely in wet weather and is laid on a solid base so that it does not subside and become uneven as time goes by.

slab paving on sand

Paving slabs come in a range of shapes and sizes, and sets are available that build up into circular features with up to three concentric rings around a central stone. Slab texture may be smooth, moulded to simulate split slate or York stone, or finished with a fine gritstone or coarser aggregate surface. Colours range from buff and grey through various shades of terracotta to red. For areas that will get only light traffic, such as patios and garden paths, the slabs can be laid on a raked and compacted sand bed about 50mm (2in) in depth. This provides continuous support for the undersides of the slabs, accommodates any unevenness in the subsoil beneath and makes it easy to get neighbouring slabs level. The sand bed can be laid directly over well-compacted subsoil, but if it is unstable or has been dug recently, a 75mm (3in) thick layer of well-rammed hardcore (broken builder's rubble) or crushed rock will be required below the sand to provide a firm base that will not subside. The joints between the slabs are filled with brushed-in sand or soil.

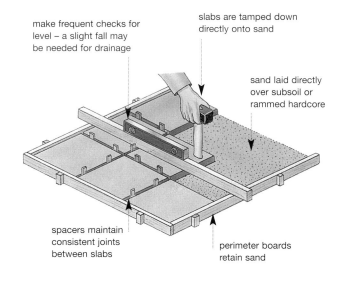

make frequent checks for level – a slight fall may be needed for drainage

slabs are tamped down directly onto sand

sand laid directly over subsoil or rammed hardcore

spacers maintain consistent joints between slabs

perimeter boards retain sand

slab paving on mortar

For areas such as drives and parking bays that have to take vehicles, slabs must be laid on a mortar bed otherwise the weight of the vehicle will make individual slabs subside and crack. The mortar bed needs a firm and stable sub-base – either a 100mm (4in) thickness of well-rammed hardcore or crushed rock, or an existing concrete drive. The slabs are laid on lines of mortar placed beneath their edges and across the centre, leaving space for the mortar to be compressed and squeezed out sideways into an almost continuous bed when the slab is placed and levelled. Joints between slabs are filled with a very dry, stiff mortar mix to avoid staining the slabs when they are placed.

Crazy paving is a form of slab paving laid using the same method as for normal paving, but with broken paving slabs as the raw material, which are laid on a continuous bed of mortar rather than mortar being applied under each one.

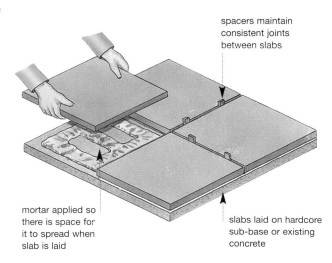

spacers maintain consistent joints between slabs

mortar applied so there is space for it to spread when slab is laid

slabs laid on hardcore sub-base or existing concrete

concrete

Concrete is a mixture of cement, sharp sand and fine aggregate – stones up to 20mm (³⁄₄in) across – which makes an ideal surface for patios, paths and drives where the emphasis is on economy and hardwearing properties, rather than a grand appearance. When formed into strips and slabs, the concrete is contained in a timber mould called formwork. This ensures that the cast slab will have clean vertical edges, and also acts as a levelling guide when pouring in the final layer of concrete. The correct mix for exposed slabs is 1 part cement to 3.5 parts combined sand and aggregate.

Once the trench has been dug and formwork established, a firm sub-base of compacted hardcore or crushed rock is laid, followed by a thin layer of sand or ballast to fill in gaps in the hardcore, with the concrete laid on top. The combined depth of the two sub-base layers should be approximately 75mm (3in) for a path or patio, and 100mm (4in) for a drive, rising to 150mm (6in) on clay soils. Similarly, the concrete layer should be 75mm (3in) thick for paths and patios, 100mm (4in) thick for drives and 150mm (6in) on clay.

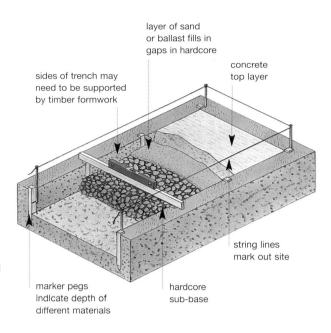

layer of sand or ballast fills in gaps in hardcore

concrete top layer

sides of trench may need to be supported by timber formwork

string lines mark out site

marker pegs indicate depth of different materials

hardcore sub-base

gravel & cobbles

Gravel is river stone typically sieved to a diameter of 20mm (³⁄₄in). It is loosely laid over a firm sub-base of well-rammed subsoil or crushed rock (hardcore is too coarse), ideally with a weed-proof membrane laid under the sub-base. A depth of 50mm (2in) is required for a garden path and 100mm (4in) for a driveway. The area will need some form of edge restraint, such as kerbstones or pegged timber boards, to prevent the gravel from transferring to adjacent lawns or flowerbeds. Cobbles are larger rounded pebbles up to 75mm (3in) in diameter. They can be laid loose but are more commonly set to around half their depth in a continuous mortar bed. They make an attractive surface but one that is relatively uncomfortable to walk on, so they are usually laid only as small feature areas within other paving or as a trim.

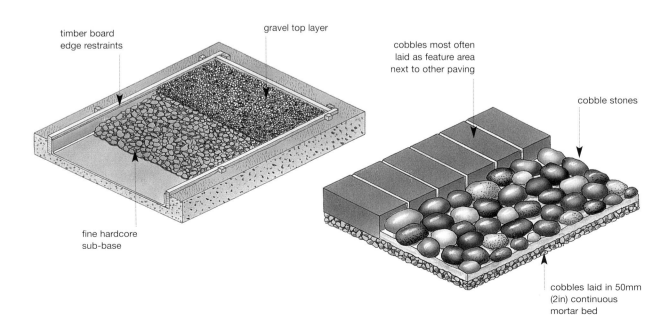

timber board edge restraints

gravel top layer

fine hardcore sub-base

cobbles most often laid as feature area next to other paving

cobble stones

cobbles laid in 50mm (2in) continuous mortar bed

planning

Whatever masonry or plastering task you intend to carry out, it is essential to be well organized before you begin. You need to decide what it is you want to achieve, and to assess its feasibility so that you know exactly what is involved in carrying it out. This will enable you to gauge whether you are able to do all the work yourself, and also to establish whether any formal consent such as Building Regulations approval or planning permission is required. The next step is to estimate what materials will be needed, and in what quantity, and to draw up checklists of the tools and any specialist equipment that will be required. Above all, you need a timescale to work to so that you can set yourself a realistic target for completion. This chapter gives an overview of some of the projects you could be considering, provides detailed information on the tools and materials you are likely to need, and examines the ways in which you can plan ahead to ensure that your renovation project meets with success.

By inserting a new partition wall, two distinct living spaces have been created from what was once a single large room.

options for change indoors

The interior of your house can be renovated in a variety of ways. You may be able to make better use of the available floor space by changing the layout of individual rooms – adding a partition wall to subdivide a large space, removing a wall between adjacent rooms to create a larger, through room, or blocking off an unwanted door opening. You may be able to get more light in dark places by fitting a new window, or by creating a new door opening or archway. A fireplace can be reinstated if it has been decommissioned, or blocked off if it is no longer wanted.

partitioning a room

Many older houses have rooms large enough to be subdivided – for example, to create a utility room off a large kitchen, or to provide a downstairs cloakroom off a wide hallway. Masonry partitions must by necessity be confined to downstairs rooms, where they can be constructed on a concrete floor or close to a supporting sleeper wall below a suspended timber one. Upstairs floors are unlikely to be strong enough to support a masonry partition without expensive strengthening of the floor joists, and a timber-framed partition will be a better option. Building a partition wall may require Building Regulations approval as far as ventilation to the new rooms is concerned, and also if either room will contain new plumbing facilities.

RIGHT *Partitioning a large room into two smaller areas is often a more practical use of space. Installing a low level partition retains the sense of openness while creating a distinct barrier.*

creating a through room

Removing the dividing wall between adjacent rooms, or creating an opening between them, can give a cluttered house a new sense of openness. The most popular conversion involves creating a larger living space from the lounge and dining room, but you could also combine the dining room and kitchen, or knock the lounge and hallway together (as long as you have a separate entrance porch). Since all these dividing walls are likely to be loadbearing, removing part or all of them will require the use of supporting steel beams called rolled steel joists (RSJs) to carry their load. The job will require Building Regulations approval, and you would be wise to call in a builder or surveyor to specify the correct size of beam and to assess the effect of the job on flanking walls and their foundations.

LEFT *The wall between the lounge and dining room has been knocked through, allowing more light into both rooms and integrating the living space area.*

installing a new window or door opening

A new window can bring much needed light to a dark room. You may want to fit an extra window in a wall that has none, to enlarge an existing opening so that you can have a bigger window, or to turn a window into French or patio doors. In an external wall, the new opening has to be bridged by a suitable lintel to carry the weight of the wall(s) above it. Indoors, a new door opening in an internal masonry wall also requires a lintel, but a much lighter duty one will suffice because the wall is thinner. Both these jobs require Building Regulations approval, and professional advice should be sought about specifying the size and type of lintel required for new openings in external walls.

altering a fireplace

Houses built with fireplaces may still have them, even if the opening of the original fireplace has been blocked off. You may decide to reinstate a blocked-off fireplace and fit a new fire surround and hearth, or to do the reverse and strip out and block off a fireplace you no longer want. The work involved for reinstating a fireplace depends on whether the old fireback was removed when it was decommissioned, while closing up involves stripping out the old fireback and blocking up the opening. You may need Building Regulations approval to reinstate a fireplace, but not for removal.

ABOVE RIGHT *Inserting an internal window helps to maximize the amount of light and creates new connections between rooms.*

RIGHT *Reinstating an old fireplace that has previously been bricked up will make an impressive feature in a room.*

creating an archway

To create an archway requires an existing or new opening with a lintel over it, and some means of creating the arch shape by filling in the top corners of the opening. The most professional way of doing this is to use arch formers in expanded metal mesh to create the sides of the arch and the curved section (called the soffit) between them. Once fixed in place, the mesh is covered with plaster to form a solid and realistic arch. Forming an arch within an exisiting opening requires no official approval.

LEFT *Adding an arch profile makes an unusual feature and will help to soften an exisiting opening.*

options for change outdoors

Outdoors, you have enormous scope for change in the way your garden is landscaped. You can build walls to mark your boundaries, subdivide the plot, retain different levels in terraces on a sloping site or create features such as arches and planters. You can link different levels of the garden with flights of steps. Brick, reconstituted stone or pierced screen walling blocks can all be used, depending on personal taste. Whether you are starting from scratch with a virgin plot around a new house, or altering what is already there, all you need are ideas and a master plan.

brick garden walls

Brick is an excellent material to use for garden walls. There is a huge range of colours, textures and finishes available, and the basic unit is light and easy to handle and lay. If you have a brick-built house, it is worth trying to find bricks for your walls that closely match the house bricks. This will give your new walls the look of having been there since the house was built. If you have never laid bricks before, start small with a simple project such as a patio planter or barbecue. As your confidence grows, you can move on to build bigger brick walls in attractive bonding patterns, adding piers and details such as soldier courses to finish off the top of the wall.

RIGHT *This fine brick wall demonstrates a version of English garden wall bond, with three courses laid as stretchers, followed by a course of alternating headers and stretchers. Weep holes along the bottom allow water to escape.*

stone block walls

Reconstituted stone walling blocks are an ideal alternative to bricks if you prefer their more natural, rustic appearance. You can use them for the same sort of structures, from simple planters to longer runs of boundary or dividing wall. They are simple for the amateur to lay because they have flat top and bottom surfaces, so level coursing is easily achievable. Some ranges have blocks of different heights, allowing you to create very natural-looking walls. The manufacturers also offer a range of matching coping stones, and it is easy to coordinate colours with other landscaping features such as paving. Stone walling blocks also make excellent retaining walls. Combined with imaginative planting, they can soon become part of the landscape.

LEFT *This stone block retaining wall cuts a swathe through the garden to create unusual planting spaces. The step design on the top of the wall keeps it at the correct height for the sloping site.*

screen block walls

Decorative pierced screen wall blocks are an attractive choice if you want to enclose an area such as a patio without having a solid wall around it. They are also excellent for concealing unwanted features of the garden such as a dustbin recess or compost bin. They have a distinctly Mediterranean appearance, and can be painted in bright colours for an up-to-the-minute look. Several different designs are available, and the manufacturers offer matching pier blocks to finish off ends, corners and T-junctions as well as coping stones and pier caps.

brick arch

A brick arch is a stunning feature, whether it is built within a boundary wall or as a free-standing structure. The secret of success is to construct some semicircular formwork to support it, and to prop this within the opening once the walls or piers at either side reach the point where the arch curve begins (the springing point). Then it is a simple matter to place the bricks one at a time, working upwards to the top of the arch before placing the single central keystone that completes the ring. Once the first ring is complete, you can add a second and even a third ring if you wish.

ABOVE RIGHT *Contrasting screen blocks with brickwork can make for a particularly elegant garden wall.*

RIGHT *This brick arch opening has been combined with a metal door for a grand and imposing side entrance to a garden.*

BELOW *Stone blocks provide the risers and slate paving slabs the treads for these delightful garden steps.*

flight of steps

Steps are an essential garden feature if you have a site with more than a gentle slope. They link different levels of the garden, and can be built into a bank or constructed as a free-standing structure against a retaining wall. The risers – the vertical sections that support the treads – can be built in brick or reconstituted blockwork, while the treads can be paving slabs, paving blocks or more bricks. The important things to remember about garden steps are that they should be shallow in step height, be wide enough for two people to pass easily, and be built in such a way that water drains freely off the treads.

tools & equipment

A variety of tools will be required to carry out the projects and repairs described in this book. Some general tools you will probably own already, such as hammers, screwdrivers, drills and saws, as well as gardening tools such as a spade, shovel and wheelbarrow. Others you will have to buy, particularly for specialist jobs like bricklaying and plastering. Buy the best tools you can afford as they will give better results than cheap tools, but remember many tools and items of access equipment can be hired, which is worth considering if you need a tool for just one job.

general tools

Examine your existing tool kit, making sure that tools are in good condition and rust-free. Look at accessories too, such as twist drill and masonry drill bits, saw blades for power saws and trimming knife blades. Some tools may be improvised from common materials: for example, a length of straight softwood batten is ideal for use as a straight edge, a sheet of board makes a good surface for mixing materials, and scrap wood can be used for pegs and packing. String will also come in handy for setting out garden building projects.

cordless drill

twist drill bits

masonry drill bits

claw hammer

rubber mallet

trimming knife

1m (3ft) spirit level

mini hacksaw

screwdrivers

mitre block

steel tape measure

handsaw

portable workbench

power tools

Apart from the cordless drill in your general tool kit, few power tools are needed for masonry and plastering projects. However, it is worth considering buying an angle grinder, which is extremely useful for cutting paving slabs to size, and a power sander, which can make light work of finishing off new plaster. A power jigsaw will also save effort when cutting wood to size for formwork and other wood-cutting jobs.

angle grinder

electric sander

jigsaw

stepladder

You will also need certain specialist tools for building work. These include bricklaying and pointing trowels, a brick jointer, pins and a string line for setting out brick courses, a brick bolster and club hammer for cutting bricks and blocks, and a metal hawk for carrying small quantities of mortar and plaster to the construction area. If you have to break up old materials, you may need a sledgehammer, pickaxe and wrecking bar. For excavating sites, mixing and moving materials and levelling bedding, raid the garden shed for a spade, shovel, wheelbarrow and rake.

bricklaying trowel

hawk

club hammer

pickaxe

pointing trowel

brick bolster

cold chisel

bricklayer's pins and string line

sledge-hammer

spade

rake

shovel

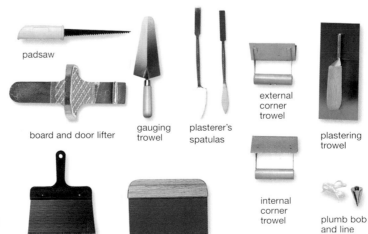

wheelbarrow

plastic bucket

brick jointer

wrecking bar

The key tool for plastering is a plasterer's trowel or steel float. Use this in conjunction with your hawk for applying plaster. Corner trowels are useful for finishing internal and external angles in plasterwork, and a gauging trowel is handy for plaster repairs. You will need a power stirrer for mixing plaster and a piece of board for holding mixed plaster close to the work. If you are working with plasterboard, you will need a panel saw, a padsaw (drywall saw), a plumb bob and line, a board lifter, a caulk blade and a taping/coating knife.

padsaw

board and door lifter

gauging trowel

plasterer's spatulas

external corner trowel

plastering trowel

internal corner trowel

plumb bob and line

power stirrer

taping/coating knife

caulk blade

construction materials – 1

You will need a range of different materials for the building projects described in this book. They are all stocked by builders' merchants, DIY superstores and garden centres, so take your time to view what they have available and to compare prices and delivery charges. Remember that trade outlets are likely to be significantly cheaper than retail ones, but may not be able to cater for small quantities. Bricks, blocks, paving materials and other masonry products are dealt with here, while cement, aggregates, additives and plaster products are covered on pages 30–1.

bricks & blocks

brick   concrete building block   thermal insulation building block

screen walling block

screen walling pier block

screen walling pier cap

decorative walling blocks

stone walling coping stones

TYPES OF BRICK AND BLOCK

• **Bricks –** Two types of brick are used for general building work: commons and facing bricks. Use common bricks anywhere appearance does not matter. Facing bricks have a decorative finish that will be exposed when built into a structure. The 'quality' of the brick describes its resistance to weathering. Internal quality bricks are for indoor use only. Ordinary quality bricks can be used out of doors, but not in severely exposed conditions or for structures such as earth-retaining walls, which will be permanently damp.

In such instances you should use special quality bricks, which are very dense and do not absorb moisture.

• **Blocks –** There are three main types of building block. Type A are dense aggregate blocks. They are loadbearing and suitable for most structural uses. Type B are lightweight loadbearing blocks, and are easier to handle than type A. Type C are lightweight non-loadbearing blocks, used mainly for internal partition walls. Lightweight blocks may be solid or hollow.

paving slabs

rustic paving slab

smooth paving slab

crazy paving slab

paving blocks

edging block – available to complement the various paving block ranges

paving block – usually rectangular but also available as squares and interlocking shapes

gravel & cobbles

gravel – stones up to around 20mm (³⁄₄in) in diameter

cobbles – stones from 50–75mm (2–3in) in diameter

SIZES & QUANTITIES

- **Bricks –** A standard brick is 215 x 102 x 65mm (8½ x 4 x 2½in), but for estimating quantities, use the working size 225 x 102 x 75mm (9 x 4 x 3in), which includes a 10mm (½in) mortar joint. You need 60 bricks per square metre (10sq ft) for a wall 102mm (4in) thick, and 120 per square metre (10sq ft) for a wall 215mm (8½in) thick.

- **Building blocks –** The standard size is 440 x 215 x 100mm (17¼ x 8½ x 4in), which is the equivalent to two bricks long and three bricks high.

- **Stone walling blocks –** Most are 100mm (4in) wide and 65mm (2½in) high – the same as a brick – and 220–440mm (8¾–17½in) long.

- **Screen walling blocks –** These are a standard 290mm (11½in) square and 90mm (3½in) thick.

- **Paving slabs –** Cast concrete slabs are generally 50mm (2in) thick, while stronger hydraulically pressed types are usually 40mm (1½in) thick. Most are square or rectangular, and come in sizes based on a 225 or 300mm (9 or 11¾in) module. Sizes range from 225mm (9in) square up to 675 x 450mm (26½ x 17¾in). You will need ten 450 x 225mm (17¾ x 9in) slabs to cover 1sq m (10sq ft).

- **Paving blocks –** A standard block is 200 x 100 x 50 or 60mm (8 x 4 x 2 or 2½in) with a coverage of 50 blocks per square metre (10sq ft). Other sizes are available, including 210 x 70mm (8¼ x 2¾in) bricks, 100mm (4in) square setts and interlocking shapes.

- **Gravel & cobbles –** For small quantities buy 25 or 50kg (55 or 110lb) bags. Larger bulk quantities will be measured by volume (in cubic metres/feet or fractions thereof) and will be delivered by your supplier.

construction materials – 2

In addition to the building materials described and illustrated on pages 28–9, you will need a number of other materials for making mortar and concrete, and for tackling plastering and repair work. All are available from the same stockists as building materials, and it pays to shop around for the best price. Remember that trade outlets usually quote prices without VAT, the addition of which can make an apparently low price increase noticeably.

cement & aggregates

cement – mixed with sand and water to make mortar, and with aggregate to make concrete

builders' (soft) sand – added to cement and water to make mortar for building purposes

concreting (sharp) sand – added to cement and aggregate to make concrete

plaster

bonding coat – water added and used as base coat on non-porous masonry

undercoat plaster – water added and used as undercoat to multi-finish plaster

multi-finish plaster – water added and used as top coat on both plaster and plasterboard

one-coat plaster – water added and used as general purpose plaster and for repair work

QUANTITIES

• **Cement –** This is usually sold in 50kg (110lb) bags, though many DIY retail outlets also sell smaller and more manageable 20, 25 and 40kg (45, 55 and 90lb) bags. The cement must be kept dry and should be used by the sell-by date on the bag.

• **Aggregates –** These are sold in 40 or 50kg (90 or 110lb) bags for small jobs, and by volume – cubic metres/feet and fractions thereof.

• **Plaster –** All types of plaster are sold in 50kg (110lb) bags, and sometimes in smaller sizes. A 50kg (110lb) bag will cover 7–8sq m (80sq ft) as undercoat, and around 25sq m (270sq ft) as a finish coat.

• **Pre-packs –** Bags of dry ready-mixed ingredients can be bought to make mortar (for bricklaying or rendering) and concrete. These are useful for small jobs and repair work, but expensive.

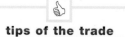

tips of the trade

You can add powdered lime or a ready-made liquid plasticizer to mortar and concrete mixes to make them more workable and help retain water, preventing over-fast drying and cracking. Other additives you may find useful are frostproofers (for use in cold weather), waterproofers (to improve weather resistance) and pigments (to colour the mix).

standard plasterboard – square edge

moisture-check plasterboard

standard plasterboard – tapered edge

vapour-check plasterboard

thermal plasterboard

fire-check plasterboard

TYPES & SIZES OF PLASTERBOARD

Plasterboard consists of a core of lightweight plaster sandwiched between outer layers of tough paper. Surfaces with a grey finish are intended to be skim-plastered, while ivory-faced boards can be decorated directly once their joints are taped. These boards may have tapered edges to allow the joints to be finished flush. Standard plasterboard or wallboard has one grey and one ivory face, and is the most widely available type. It comes in standard 2.4 x 1.2m (8 x 4ft) sheets and in smaller sizes, in both 9.5 and 12.5mm (3/8 and 1/2in) thicknesses. Boards containing a vapour barrier are available to prevent condensation forming behind them, or with expanded foam insulation bonded to the rear face for use in dry lining work. Other types of board available include moisture- and fire-check varieties.

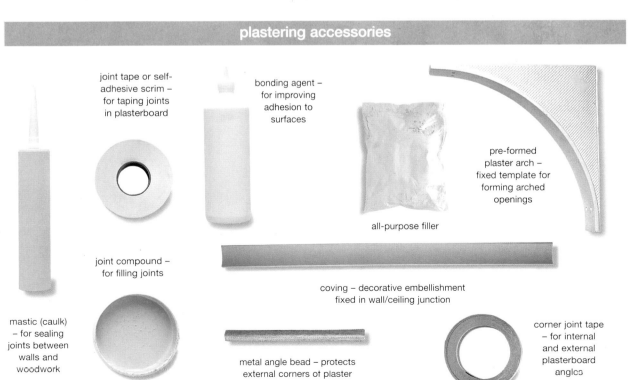

joint tape or self-adhesive scrim – for taping joints in plasterboard

bonding agent – for improving adhesion to surfaces

pre-formed plaster arch – fixed template for forming arched openings

all-purpose filler

joint compound – for filling joints

coving – decorative embellishment fixed in wall/ceiling junction

mastic (caulk) – for sealing joints between walls and woodwork

metal angle bead – protects external corners of plaster

corner joint tape – for internal and external plasterboard angles

how to start

Before you can contemplate carrying out any project, you need to be fully prepared. This not only involves such basics as estimating quantities of materials, deciding on the right order of work and planning your schedule, but also means checking whether the work needs any formal consent from your local authority and, if it does, applying for planning permission or Building Regulations approval as appropriate.

making a scale drawing

Most construction jobs will proceed more smoothly if you prepare scale drawings before you start. They do not have to be done to architectural standards but should be accurately measured and recorded. Bear in mind the following guidelines when making your drawing:

- use graph paper rather than plain paper

- invest in some good pencils and fine-line drawing pens, a ruler and square

- work to a sensible scale – 1:20 is ideal for most projects – and use metric measurements for accuracy

- draw the ground plan and elevations if appropriate, and use them as a guide for estimating materials

estimating materials

If you are doing a project yourself, your labour comes free. It is the materials that cost the money, so you should set aside some time to estimate how much you need of all the materials that the job requires. Use the following guidelines to help you prepare your shopping list:

- make sure you know the dimensions of whatever solid materials you will be using – bricks, blocks, paving slabs, plasterboard and the like – and use these as units to count the required quantities

- with bagged and loose materials such as cement, sand and plaster, find out the coverage from the bag or your supplier and use this to work out how many bags or what volume of loose material you will need

- once you have an itemized shopping list, compare prices and then use the list as your order form at your chosen supplier

- do not be tempted to overestimate – it is wasteful to go beyond a sensible safety margin of, say, 5 per cent

- do not underestimate – this is annoying if you run out of something just as the shops are closing

planning the order of work

For all but the smallest projects, breaking the job down into distinct stages will not only help you to carry them out in the correct order, but will also make it easier for you to chart your progress. The factors outlined below must be taken into account to ensure that your project is carried out as efficiently and successfully as possible:

- check that you have all the tools and materials needed for each stage

- estimate the time required to complete each stage to your satisfaction

- never underestimate the time DIY work takes – there are always hidden delays, unforeseen problems and other hold-ups that can wreck a schedule, especially if you are carrying out the work in the evenings and at weekends

- take into account the disruption the work will cause to the running of the household

planning permission

Some building and alteration work in the home needs formal consent from your local authority. If in any doubt, always check before starting work – if approval was needed and you proceeded without it, you may receive an enforcement notice or a fine should the infringement be discovered within 12 months of its completion. A general outline of when planning permission will be required, and how to apply for it, is given below:

- planning control is mainly concerned with changes to the way the house looks and is used, so you are unlikely to need planning permission for most of the projects described in this book unless internal alterations lead to a change of use – for example, a room being converted into an office, or the creation of a granny flat

- planning permission will also be required for alterations to buildings within conservation areas

- all local authorities have planning officers who will be able to tell you if any work will require planning consent

- many authorities publish booklets outlining the scheme that may prove helpful in determining if you need approval – many of these are available free of charge from local libraries

- in cases where you do need planning consent, you will have to pay a fee and submit plans, which you must file with the local authority along with an application form

- you can submit the application yourself or you can get someone to do it for you, such as an architect or architectural technician – the advantage of doing this is that they will be familiar with the process and will know what is required, and they will also be able to prepare any necessary drawings

- by law, once the application has been made the local authority has to give you a decision within eight weeks

- until your application has been approved, you must not start work or you could be in trouble

- planning consent is valid for five years – if you do not start the work within this time, you will have to make another application

building regulations

Building Regulations are concerned with local and national building codes and specify the types of materials that can be used, minimum dimensions and other essential requirements that must be complied with. Although you may not need planning permission, any major alterations to a home will be subject to the Building Regulations outlined below:

- building control is concerned with structural safety and the health of the property's occupants, so many of the indoor projects in this book will need Building Regulations approval for any structural alterations being made

- outdoor projects are in general not subject to Building Regulations control, but always check with your local authority before starting work

- responsibility for Building Regulations falls to a Building Control Officer (BCO) employed by the local authority, so ask his or her advice before you start work – although the officer will be unable to recommend a builder, he or she may have one or two ideas that you could incorporate into your scheme

- once work has started, it is more than likely that the BCO will visit the site to ensure that all work is being carried out in line with the necessary regulations

- if the project is a large one, the BCO will visit the site at different phases of the project to ensure that all is well – it is up to you to ensure that you notify him or her at the correct time

- at the end of the job, the BCO will issue a completion certificate to certify that the work has been completed in accordance with Building Regulations

- if any work is carried out in contravention of the Building Regulations or planning laws, you could be faced with a large bill to put the matter right or forced to restore the work to its previous condition

- If all this sounds rather daunting, it need not be; simply contact your local planning office and ask their advice– they would much rather that you contacted them at an early stage as this could save work on both sides

dealing with professionals

When planning a project, it is important to be aware of your limitations and to be prepared to leave some jobs to the professionals. You may lack the particular skill to do a job well. You may need extra manpower to undertake certain major tasks. You may simply realize that a job will take too long to complete if you are tackling it single-handed on a part-time basis. Whatever the reason, it is important to know how to get the right help, and how to manage it when you have found it.

finding a contractor

For jobs that require professional help, your first step is to track down the type of contractor you need. The list below outlines the various methods for doing this:

- the best way is by personal recommendation – asking friends or neighbours about people they have employed – so that you can find out about the contractor's workmanship and reliability, and also inspect the work they have done

- if no one you know has employed the type of professional you are seeking, look around your neighbourhood for properties where work of the type you want is being carried out – many contractors put up a sign or park their vans outside houses they are working on, and most homeowners are only too happy to talk about their home improvement projects and recommend their contractors if they are doing a good job

- telephone directories and local newspapers are your next port of call – Yellow Pages lists contractors by trades

- display advertisements are useful because they usually indicate whether the person or company is a member of a relevant trade organization, but remember that such claims can be bogus, so always check with the organization concerned before proceeding

- ask firms that you contact by telephone to put you in touch with local customers for whom they have worked so that you can call and enquire about them personally

- trade organizations (see page 143 for details) will often supply a list of members working in your area, and many also offer a complaints service and an arbitration scheme in the event of a dispute

inviting quotations

Once you have found a contractor who appears interested in carrying out the work you want done, arrange an initial site meeting so that you can explain what is involved in more detail. The process outlined below provides guidance on how to deal with obtaining, checking and accepting a quotation:

- offer copies of any plans or drawings you have made, and invite the contractor to submit a quotation

- always obtain two or three quotations for any professional job

- ask each contractor how long the job will take, how long the quoted price will remain valid and whether the price includes VAT

- remember that a quotation (or estimate) is just that, and the price may alter if you change your mind about materials or details of the work – you will get a fixed price only for relatively straightforward jobs with little scope for variation

- read quotations carefully to make sure the contractors have quoted for everything you want, including the use of particular products – high estimates may be a contractor's way of telling you that he or she does not want the job, while low ones may conceal the fact that the contractor will be using inferior materials or subcontracting work to inexperienced and low-paid people

- when you have chosen a contractor, write accepting the quotation and ask when work can start, when it is scheduled to finish, how payments are to be made and whether the firm has employer's liability and public liability insurance – these cover damage done to the property and injury caused to employees or third parties

If you are intending to hire professionals to carry out major construction work on your home, it is vital to ensure that you pick a competent and reputable individual or firm and that a detailed costing and timescale are agreed upon in advance.

when work starts

Do not make the mistake of paying for work before it starts. On small projects taking no more than a couple of weeks, reliable contractors will expect to be paid when the job is complete. On longer, more complex projects it is normal to stage payments, especially if you are using expensive materials that have to be bought at the outset. The points below will help you to deal with such issues in the best way:

- agree how payments will be made before work begins, and retain the largest payment until the job is finished

- give clear instructions and specifications for any work, outlining precisely what you require the contractor to do and what is to be done by others (you may plan to do some of the work yourself, for example) – this avoids misunderstandings and disputes later on

- as work progresses, keep out of the contractor's way during the day if at all possible – no one likes being watched as they work, so wait until the contractor has left for the day to inspect the work

- have a brief meeting at the start or end of each day to discuss progress, agree on any variations to the original plans in writing, and iron out any complaints you may have about workmanship or finish – most things can be sorted out by a process of compromise

- pay up on time – contractors have bills to pay, too – and recommend them to your friends if you feel they have done a good job at a fair price

- only threaten legal action if things remain unresolved to your satisfaction once the work has been completed

safety at work

Thousands of people are injured every year and more than a few are killed while carrying out home improvement projects. Many of these accidents could have been prevented if the right safety equipment had been used and, above all, if access equipment had been employed safely. The other major cause of accidents is simple carelessness – with power tools, with sharp-bladed hand tools and with certain DIY liquids and powders that can burn skin, injure eyes or give off inflammable or choking vapours.

safety equipment

There is a wide range of DIY safety equipment available to protect vulnerable parts of the body from hazards such as dust, noise, fumes, lead in old paint, flying or falling debris, coarse building materials and chemicals. Other items such as work boots and disposable overalls simply make DIY work more comfortable.

protective gloves

work boots

Be prepared to patch up minor cuts and scratches by keeping a first aid kit in your workshop. It should contain assorted plasters, sterile gauze pads, antiseptic cream, scissors, tweezers for removing splinters and other embedded objects, and perhaps an eye bath for rinsing out dust and splashes with clean water.

lead test kit

respirator mask

hard hat

knee pads

dust mask

goggles

ear defenders

first aid kit

tool safety

The most dangerous hand tools are those with blades or sharp points. When using one, make sure that you keep your hands behind the direction of operation of the tool so that if it slips you will not be injured. Be especially careful when using trimming knives, which cause more accidents than any other tool. Keep blades sharp – you are more likely to be injured using a blunt tool because you will have to force it to make it cut properly, and that in turn can cause the tool to slip. Use an oilstone for sharpening chisels and plane blades,

and replace blades on knives and similar tools as soon as they become blunt. Store bladed tools such as chisels, knives and saws with their blade guards on.

Treat power tools with care. Read the operating and safety instructions before you use one for the first time, and carry out a spot check every time you use the tool to make sure that its casing, flex and plug are in good condition. If they are damaged, get the tool repaired before using it again. Always double-check that blades, drill bits and other accessories are

properly fitted and completely secure before switching the tool on. Never try to bypass or deactivate any safety guard. Avoid wearing loose clothing that could be caught in moving parts. Wear the appropriate safety equipment when drilling, sawing or sanding.

Read the instructions with all DIY materials before using them. Keep packaging or separate instruction leaflets in a file for future reference. If the print is too small to read clearly, boycott the product and complain to the manufacturer.

You will not need tall ladders for the projects contained in this book, but you will still need low-level access equipment for some of them – including stepladders, multi-purpose ladders and low-level working platforms. All need to be set up properly and used safely if accidents are to be avoided. Make sure that any equipment you buy – or hire – meets the relevant national safety standards and is in good condition before you start to use it.

stepladders

Check that the stepladder is opened fully and locked in that position. If it is a multi-way design, make sure it is locked in the correct configuration for the job you are doing. Stand the stepladder on a firm and level base, and position it front-on to the work, not sideways. When you are on the stepladder, keep a grip with one hand if possible while you work. Otherwise, brace your knees against the treads for support. Keep both feet on the treads. Do not work with one foot on

another nearby surface in case the stepladder becomes unstable and causes a fall. Never over-reach – get down and reposition the stepladder if necessary, or use a taller stepladder. Do not keep loose tools on the top tread or platform, from where they could fall and cause injury. Never have more than one person on the stepladder at any time.

working platforms

If you need same-height access to a large area of wall, stepladders are a nuisance because you have to keep repositioning them. A better solution is to hire some adjustable steel trestles and scaffold boards, and use them to set up a low-level work platform spanning the work site. Choose the trestles to suit the platform height you require – different sizes offer working heights from 500mm (20in) above floor level upwards. You need a trestle every 1.5m (5ft). Make sure they are all standing

level and square to each other, and fit the boards so that they butt tightly against one another.

stand stepladders on a firm base front-on to the work

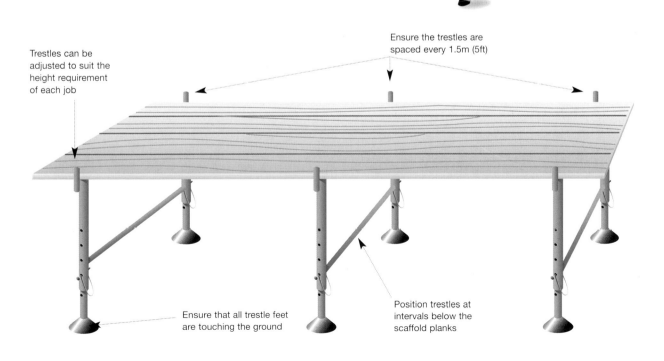

Trestles can be adjusted to suit the height requirement of each job

Ensure the trestles are spaced every 1.5m (5ft)

Ensure that all trestle feet are touching the ground

Position trestles at intervals below the scaffold planks

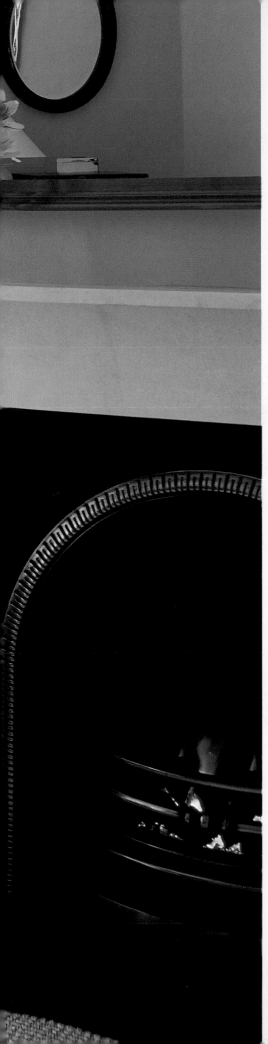

making changes indoors

The projects in this section involve internal alterations that will change the layout of your home, alter its traffic routes by creating new door openings and blocking up existing ones, and improve its amenities by creating new window and door openings and reopening or decommissioning fireplaces. Most of these jobs are straightforward projects, well within the scope of a competent DIY enthusiast. Others involve major structural alterations that will need Building Regulations approval and careful workmanship to ensure that they do not compromise the house structure in any way. Unless you have tackled such work before and are confident of your ability to carry out the job correctly and safely, you would be well advised to call in professional help.

Opening up a disused fireplace can add
an impressive focal point for a room,
whether or not you install a real fire.

building a block partition wall ⁄⁄⁄

If you need to sub-divide a ground-floor room, a masonry partition wall built using lightweight blockwork is the best solution. It creates a substantial wall with excellent soundproofing qualities, and provides a structure to which you can fix wall units or other fittings at any point. It can be built directly off a concrete ground floor, but timber ground floors will need reinforcement to take the weight of the structure (see tips of the trade below).

tips of the trade

If the floor is timber, lift floorboards to gain access to the underfloor void. Build a sleeper wall of honeycomb brickwork on the oversite concrete to support the joists that will carry the weight of the wall. Allow it to harden for 48 hours before starting to build the new partition wall.

tools for the job

plumb bob & line

pencil

tape measure

spirit level

cordless drill

masonry drill bit

spanner or screwdriver

hammer

string line & masonry nails

bricklaying trowel

gauging trowel

club hammer

brick bolster

safety goggles

gloves

1 Use a plumb line and bob and a pencil to mark a vertical line on the wall against which the new wall will be built, to indicate the position of the metal wall extension profiles that will tie the new wall to the existing one. Hold the profile strip against the wall, checking that it is vertical with a spirit level, then mark and drill a hole at each fixing position, using the size of masonry drill bit recommended by the profile manufacturer.

2 Insert a fixing in each hole and tighten. The fixings supplied with profiles vary, and may be a coach screw and washer or a large-diameter screw. Use a spanner to tighten the former, and a screwdriver for the latter. Check with a spirit level that the profile is vertical, and add cardboard or hardboard packing behind hollow spots. Repeat the process to fix another profile to the opposite wall, in line with the first one. Draw a line on the floor to connect the centre of the profiles.

3 Place a block next to the first profile so that it overlaps the profile by equal amounts each side. Mark the position of one top corner

on the wall, remove the block and nail a string line to the wall at the marked point. Run it across the room and fix it to a nail placed at the equivalent position next to the other profile.

4 Use a bricklaying trowel to spread a strip of mortar on the floor next to the first profile. It should be a little longer and wider than the block you are using, and thick enough to bed down to about 10mm (½in) when the block is placed and tamped down into it.

5 Use a gauging trowel to spread a cap of mortar 50mm (2in) thick on one end of the first block – a process known as buttering.

6 Turn the block to the horizontal and set it in place on its mortar bed, pushing it tight up against the profile so that excess mortar is squeezed out of the joint. Check that the face of the block is correctly aligned with the string line and adjust the block's position if necessary.

7 Tap the block down with the end of the trowel handle to settle it into its mortar bed. This will squeeze out excess mortar, leaving a bed joint about 10mm (½in) thick. Check that the block overlaps the profile by equal amounts at each side and adjust if necessary.

8 Use a spirit level to check that the block is sitting level and that its face is vertical. Tap the block down into its mortar bed as necessary. Repeat steps 4 to 8 to lay more blocks one by one across the room. Pick up excess mortar from the bed joint as you work and use it to bed the next block. If you need to cut a block to size to complete the first course, cut it as in step 10 and set it in place against the profile.

9 With the first course completed, spread mortar on top of the first block about 20mm (¾in) thick. Hook on one of the wall ties provided with the profile and bed it into the mortar. This will tie the new structure to the profile as the wall rises.

10 Start the second course with a half block. Mark the cutting line on the face of the block and cut it with a brick bolster and club hammer. Cut a groove along the line first, then reposition the bolster at the centre of the line and strike it harder to split the block.

11 Butter mortar onto the block and position it against the profile, as in steps 5 and 6. Reposition the string line level with the top edge. Continue the second course in whole blocks, cutting the final block to size if necessary. You will need to fit a tie profile at the opposite wall as well. Do not lay more than five courses in a day or the weight of the blocks will force mortar out of the bed joints. Cut blocks for the final course to the width required to fit the gap beneath the ceiling.

DOOR OPENINGS

Mark the position of door openings on the floor, ideally a whole number of block lengths away from the first wall, and lay blocks up to the first mark. Measure the width of the door frame, and place the next block at the other side of the opening. Complete the first course, then build up the rest of the wall to the height of the door frame. Set a lintel across the opening (see pages 44–5) and complete the wall.

removing a masonry wall ⁄⁄⁄⁄

Before you can contemplate removing part or all of a masonry wall, it is essential to get professional advice from a builder or surveyor. The first step is to establish whether the wall is loadbearing – that is, carrying the weight of floor joists and another wall in the room above. The second is to calculate the type and size of beam needed to support that load once the wall is removed. The job requires Building Regulations approval from your local authority.

It is unusual to remove an entire wall because this can affect the stability of the flanking walls at right angles to it. Instead, a pier of masonry is left at each side of the room to support the ends of the beam – usually an I-section rolled steel joist (RSJ) if the wall is loadbearing. This beam is generally fitted with its top edge at

ceiling level, supporting the floor joists, but can be installed lower down. Either way, the first step is to remove plaster from the wall along the line of the beam to expose the wall's structure and reveal whether it is built of brick or block. When the level of the RSJ has been decided, the second step is to cut out bricks

or blocks and create holes in the wall through which you can insert stout timber supports called needles. These are then jacked up against the ceiling (or the tops of the holes) using adjustable steel props. When this supporting structure is in position, you can begin the job of inserting the RSJ and demolishing the wall.

making the opening

Clear the area on either side of the wall of furniture and floor coverings. Stack furniture under dustsheets and roll carpets or other sheet floor coverings out of

the way. Make sure you have all the tools, equipment and materials necessary for the job in the room, then close the doors to keep dust out of the rest of the house.

1 Mark the opening on the wall surface and chop away plaster down each side of the opening to expose the bricks or blocks. If the opening size is not critical, adjust its position to coincide with the edges of bricks or blocks.

2 Cut holes through the wall at the chosen level to support the needles. If the opening does not coincide with vertical joints in the masonry, hire a power tool called a wall cutter to make the opening. It will cut through walls up to 250mm (10in) thick.

3 Insert the needles and position a prop under each end. Set the ends of the props on a scaffold board to spread the load on a timber floor.

4 Tighten the props up to support the needles.

5 Start chopping out bricks or blocks at the top of the new opening. Work down the wall course by course, cutting downwards through bricks or blocks that project into the opening. Prize off and save skirting boards.

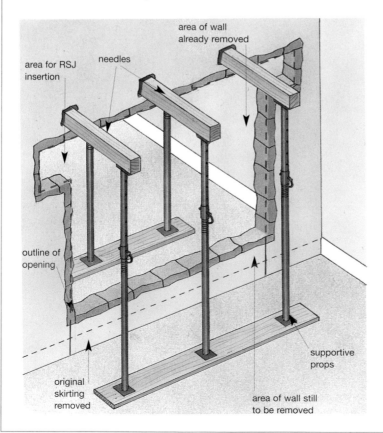

area of wall already removed

area for RSJ insertion

needles

outline of opening

original skirting removed

supportive props

area of wall still to be removed

You will definitely need help to lift the RSJ into position – two extra people are ideal. Manoeuvre the RSJ into the room and place it next to the wall. Cut out bricks or blocks at each end to create a bearing for the ends of the RSJ. Insert a concrete padstone at each side if this is recommended by your professional adviser, then check all your measurements before proceeding.

1 Lift the RSJ to the level of its bearings and slide it into place. Make sure the stepladders or supports you are using are strong enough to take the weight.

2 Drive wooden wedges between the bearings and the underside of the RSJ to force it up tightly against the masonry or ceiling joists that it will be supporting. Check that the RSJ is level and overlapping its bearings by the same amount at each end.

3 Pack mortar into the gap between the bearing and the underside of the RSJ at each end. Insert slate packing into the mortar if it is thicker than 25mm (1in).

4 Reposition three of the props – one near each end and one in the centre – to support the RSJ. Take out the needles and make good the holes.

5 Make good the sides of the opening and plaster them. After 48 hours, remove the props and box in the RSJ with plasterboard to protect it from fire (see below for more details).

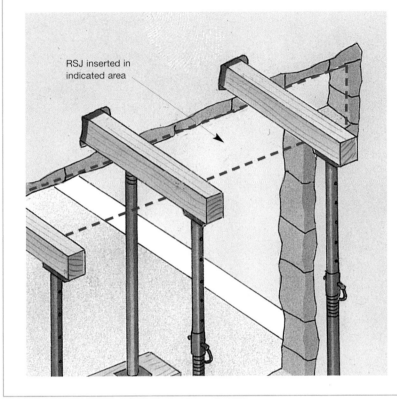

RSJ inserted in indicated area

43

POINTS TO REMEMBER

There are several important points about procedure and safety to bear in mind when removing a wall. Ignore them at your peril.

• **Needles and props** – The size and spacing of needles and props depend on the load they are supporting, so ask your professional building adviser for details. Needles should be placed beneath the centre of a brick or block, not beneath a joint. Adjustable steel props can be hired and come in four sizes. Choose the size that matches your ceiling height most closely when partly extended. If the room in which you are working has a suspended timber floor, hire scaffold boards on which to stand the props and spread the load.

• **Rerouting services** – If there is a radiator on the wall you plan to remove, disconnect it and reroute the pipework to the new radiator position before starting demolition work. Disconnect and reroute cables to light switches or power points on the affected wall.

• **Boxing in RSJs** – To protect an RSJ in the event of fire, it must be completely encased in plasterboard. Drive shaped wooden wedges into the sides of the RSJ, then nail timber battens to these to form a framework for the plasterboard. Nail the boards to the battens and apply a skim coat of finish plaster to complete the job.

• **Safety equipment** – Wear a hard hat, safety goggles, a dust mask, work gloves and stout boots when demolishing the wall, to protect you from falling debris and the large amounts of dust the work creates. Clear the rubble as you go – ideally through a window or an external door in one of the rooms if there is one. Take it directly to a skip so that you do not have to carry or barrow it through the rest of the house.

adding an interior door opening ⁄⁄⁄⁄

Creating an opening in a masonry wall for a new doorway is a scaled-down version of removing a wall (pages 42–3). The masonry above the new opening must be supported with a lintel, whether the wall is loadbearing or not. However, unless the opening is wider than a standard doorway, you will not need professional advice in choosing and sizing the lintel, though you will still need Building Regulations approval before proceeding with the project.

Start by deciding on the size and approximate position of the new door opening, then buy the door and frame so that you can use the frame as a template for marking the opening on the wall surface. Cut away the plaster along the top and down the sides of the opening to expose the masonry below, and adjust the position of the frame so that it coincides with the vertical joints between the bricks or blocks at one side of the opening. This will minimize the amount of cutting needed to create the opening. At the same time, clear plaster above the opening so that you can locate and chop out two bricks or blocks and insert timber needles to carry the weight of the wall while you cut out the opening. These should be located at least one block or three bricks higher than the proposed top of the door opening. Once the needles are in place and have been securely propped (see page 42), you can begin the job of creating the opening and inserting the lintel.

making the opening

Clear the area on either side of the wall of furniture, roll back carpets and protect other fixed floor coverings with dustsheets. Collect all the tools and materials you need for the job in the room, then close other doors to stop dust from spreading through the house.

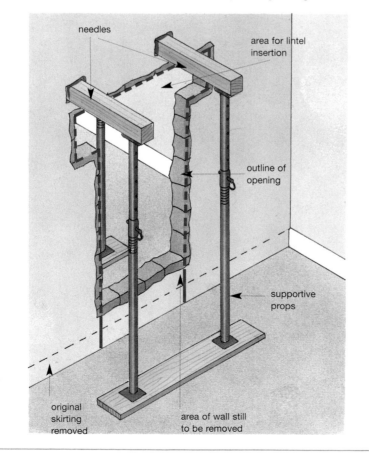

needles

area for lintel insertion

outline of opening

supportive props

original skirting removed

area of wall still to be removed

1 Use the door frame as a template to mark the outline of the opening on the wall. Chop away plaster around the outline on one side of the wall to find the vertical joints between the bricks or blocks. If the door position is not critical, reposition the frame so that one side coincides with these joints.

2 Cut out bricks or blocks to create holes for the needles, one above each side of the proposed opening.

3 Insert the needles and position a prop under each end. Set the ends of the props on a scaffold board to spread the load on a timber floor.

4 Tighten the props up to support the needles.

5 Cut out bricks or blocks across the top of the opening using a brick bolster and club hammer. Then work down the wall course by course, cutting downwards through any bricks or blocks that project into the opening. Remove the skirting board and continue removing masonry down to floor level.

Unless the door opening you are creating is wider than a standard one, you can install the new steel lintel single-handed. However, an extra pair of hands is useful in positioning the lintel accurately.

You need a lintel 1050mm (42in) long to bridge a standard internal door opening. Ask your supplier for advice on the correct type to use, depending on whether the wall is brick or blockwork.

1 Cut out bricks or blocks at the corners of the opening to accept the ends of the lintel, then prop the frame in the opening and wedge it in place down the sides.

2 Lift the lintel into place on top of the frame, making sure that it overlaps its bearings by the same amount each end.

3 Pack cut pieces of brick or block beneath the lintel's ends to support it so that it clears the top of the frame by 3mm ($\frac{1}{8}$in). Check that it is level, then fill the cut-outs at each end with mortar.

4 If necessary, use cut bricks or blocks to fill any gap between the top of the lintel and the masonry above. Allow to set for 24 hours, then remove the needles and make good the holes.

5 Secure the frame using three frame fixings into the masonry at each side. Fill any gaps between frame and masonry with mortar or foam filler.

6 Make good the plaster around the frame, ready for the door to be hung and the architrave to be fitted around both sides of the opening.

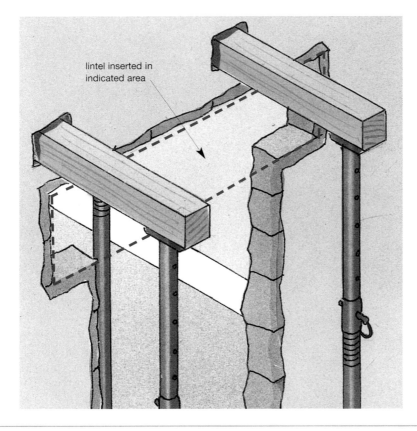

lintel inserted in indicated area

45

POINTS TO REMEMBER

There are several important points about procedure and safety to bear in mind when creating a new door opening.

● **Needles and props** – You will need two needles and four props. Needles should be placed beneath the centre of a brick or block, not beneath a joint. Adjustable steel props can be hired. Choose the size that matches your frame height most closely when partly extended. If the room in which you are working has a suspended timber floor, hire scaffold boards on which to stand the props and spread the load.

● **Lintel types** – Several patterns of lintel are manufactured for bridging door openings. Steel lintels in corrugated or box section are the easiest to install. A 100 x 65mm (4 x 2$\frac{1}{2}$in) concrete lintel is an alternative but is much heavier. Ask your building materials supplier for advice once you have established what type of masonry the wall is built in.

● **Rerouting services** – If there is a radiator on the wall you plan to remove, disconnect it and reroute the pipework to the new radiator position before starting the demolition work. Similarly,

disconnect and reroute cables to light switches or power points on the affected wall.

● **Safety equipment** – Wear a hard hat, safety goggles, a dust mask, work gloves and stout boots when creating the opening, to protect you from falling debris and the large amounts of dust the work creates. Clear the rubble as you work – ideally through a window or an external door in one of the rooms if there is one. Take it directly to a skip so that you do not have to carry or barrow it through the rest of the house.

making an external opening ⁄⁄⁄⁄

Installing a window or external door where none exists is a more complex task than making an opening in an internal wall because of the extra thickness of the external wall. You need to find out whether your house has been constructed using solid or cavity walls so that the correct type of lintel can be used. For this you will need to seek professional advice from a builder or surveyor because it is beyond the capabilities of all but the most experienced amateur.

Before you can install a new window or door opening, check with your local authority to see whether there are any restrictions concerning its positioning. The window might not be allowed if it overlooks windows in a neighbour's house, or if it significantly alters the appearance of your house – especially if it is a historic building or is sited in a conservation area. If you get the go-ahead to proceed, you will also have to apply for Building Regulations approval for the work.

Choose the style of the new window or door carefully so that it matches existing ones as closely as possible in style and proportions. A wide range of standard window sizes is available in a variety of styles. Alternatively you could have one made to measure by a carpenter, but this would be more expensive than an off-the-shelf window.

If the new window opening is upstairs, you will need scaffolding or a slot-together access tower outside the house to act as a work platform.

making the opening

Unless your house has solid walls, you will have to make the opening in two stages. Locate it so that the sides of the frame align with vertical joints in the brickwork if possible. Mark the window or door dimensions on the wall and drill holes through it to the inside with a long masonry drill bit. This transfers the outline to the inner face of the wall. Join up the four drill holes with pencil lines.

1 Set up adjustable props with a scaffold board above them to support the ceiling above the opening. Stand the feet of the props on another board if the floor is a suspended timber type.

2 Cut away the plaster across the top of the proposed opening and down the sides, working to the pencil outline. Chop out the course of bricks or blocks across the top of the opening.

3 Work down the wall course by course, cutting downwards through bricks or blocks that project into the opening. Continue until you reach the bottom of the opening for a window, or floor level for a door. Remove any cavity insulation.

4 Fit two or three wall supports within the opening in the inner leaf of the wall, with their horizontal blades hammered into the mortar joint in the outer leaf that will form the top of the opening. Fit an adjustable steel prop between each support and the sill, and tighten fully.

5 Chop out the brickwork course by course until the opening is clear.

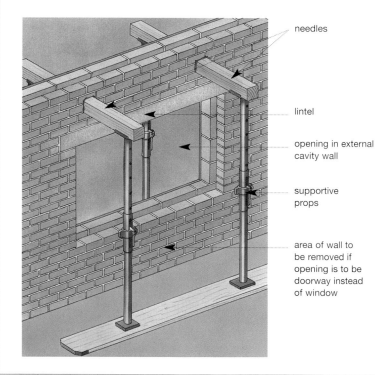

needles

lintel

opening in external cavity wall

supportive props

area of wall to be removed if opening is to be doorway instead of window

You can use one of two types of steel lintel to support a cavity wall. One is shaped like an inverted U in cross section, while the other has the cross section of a slim right-angled triangle. The U-shaped or triangular part fits inside the cavity, and the flat base of each lintel rests on the brickwork at either side of the opening. Both are packed with polystyrene insulation before being installed, to prevent them from acting as cold bridges and causing condensation.

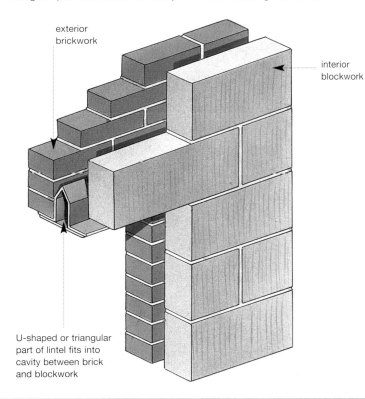

exterior brickwork

interior blockwork

U-shaped or triangular part of lintel fits into cavity between brick and blockwork

Openings in solid external walls should be bridged with a steel box lintel. The method of installation is the same.

1 Chop out a slot at each side of the new opening to provide bearings at least 150mm (6in) wide for the ends of the lintel.

2 Remove the wall supports and lift the lintel into place. Check that it is level and is overlapping the bearings by the same amount at each end.

3 Wedge the lintel at each end so that it touches the brick or blockwork above the opening. Fit two adjustable props within the opening to support it.

4 Pack mortar into the gap between the bearing and lintel, and allow to harden for 24 hours. Insert slate packing if the gap is more than about 25mm (1in).

5 Remove the props and make good at the top and sides of the opening, inside and out, ready for the new window or door frame to be installed.

POINTS TO REMEMBER

There are several important points about procedure and safety to bear in mind when creating a new opening in an external wall.

• **Supports and props** – Wall supports should be placed beneath the centre of a brick or block, not beneath a joint. Adjustable steel props can be hired and come in four sizes. Choose the size that matches your ceiling height most closely when partly extended. If the room in which you are working has a suspended timber floor, hire scaffold boards on which to stand the props and spread the load. Use another board across the sill on which to stand the props holding the wall supports in position.

• **Closing the cavity** – Fit vertical damp-proof courses (dpcs) against the inner face of the outer leaf at each side of the opening before blocking the cavity with cut pieces of walling block or a proprietary cavity closer. Fit a horizontal dpc across the sill before installing the window or door frame.

• **Rerouting services** – If there is a radiator on the wall you plan to remove, disconnect it and reroute the pipework to the new radiator position before starting the demolition work. Similarly, disconnect and reroute cables to light switches or power points on the affected wall.

• **Safety equipment** – Wear a hard hat, safety goggles, a dust mask, work gloves and stout boots when breaking out the opening, to protect you from falling debris and the large amounts of dust the work creates. Clear the rubble from indoors as you work so that it does not pose a trip hazard. When breaking through the outer leaf of the wall, drop the rubble outside.

closing up a doorway ⚏

If you have altered the layout of your house by creating a through room or opening up a new doorway, you may well have an existing door opening that is no longer required. Closing it off will restore valuable wall space in both of the rooms that the door connects. Unless the wall is a timber-framed partition, the best way of filling the opening is to use lightweight blockwork. With care, the result will be an invisible repair.

tools for the job

screwdriver
wrecking bar
cordless drill/driver
masonry drill bit
bricklaying trowel
gauging trowel
spirit level
club hammer
brick bolster
hawk
plasterer's trowel
safety goggles
gloves

1 Begin by unscrewing the door from its hinges (or the hinges from the frame, whichever is easier) and set it aside. Prize off the architraves by inserting the straight end of a wrecking bar between one side of the door frame and masonry and levering it away. It will probably have been fixed with cut nails, which will tear through the wood as you lever out the frame. Repeat on the other side. The head will fall away as you remove the other side.

2 It is important to tie the new blockwork to the existing masonry for strength and stability. You can buy wall ties designed specifically for this job. They should be inserted at intervals up each side of the opening to fit into the mortar course between the infill blocks. With standard 215mm (8½in) high blocks, you need to drill the holes at 225mm (9in) spacings to allow for the 10mm (½in) thick mortar joint between them.

3 Push a wall plug into each hole and screw in a tie as far as it will go by hand. Insert a screwdriver between the arms of the tie and use it as a lever to give the tie the last few turns needed for a secure fixing.

4 Mix some mortar and use a bricklaying trowel to spread a bed of it across the opening. The strip should be a little wider than the block and thick enough to bed down to a joint about 10mm (½in) thick.

5 Butter a cone of mortar onto the end of the first block with a gauging trowel. If it keeps falling off, make the mortar a little wetter. You need enough to form a 10mm (½in) thick joint when it is compressed.

6 Place the first block on the mortar bed with its end beneath the first wall tie and force the buttered end against the existing masonry. Tamp it down and use a spirit level to check that it is both level and plumb (with its outer faces vertical). Complete the first course with a second block cut to the appropriate length. Complete the remaining courses, filling the gap between the last course and the lintel with blocks cut to the required height. Plaster the new wall (see right, steps 4–7).

POINTS TO REMEMBER

• **Cutting blocks –** Mark the cutting line on the face of the block, then use a brick bolster and club hammer to cut a shallow groove along it. Position the bolster at the centre of the groove and strike it harder to split the block. Wear goggles to protect your eyes.

• **Timber floors –** If the floor running through the opening is a suspended timber type, screw a length of 100 x 50mm (4 x 2in) sawn timber to the floor across the threshold to act as a sole plate for the infill blockwork. This will eliminate any risk of planks moving and causing cracking in the infill.

• **Ruling off –** After applying the finish coat of plaster, hold a timber batten on edge across the repair at floor level, with its ends resting on the existing plaster at each side. Move it up the wall with a side-to-side motion to rule (scrape) off any excess plaster. Use a wet plasterer's trowel to give the plaster a smooth finish. If the joint between old and new is still visible, smooth it with an orbital sander.

filling the opening – alternative method

If the existing wall is built in blockwork, an alternative to using frame ties is to remove the cut blocks in alternate courses at each side of the opening. You can then bond the new infill blocks into the structure as if they had always been part of the wall. You will still need a cut block to complete each course of the infill. Alternate the position of the cut blocks to left or right in subsequent courses to prevent the vertical joints from lining up.

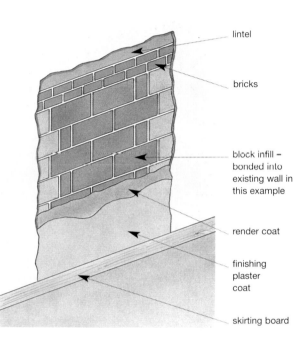

lintel

bricks

block infill – bonded into existing wall in this example

render coat

finishing plaster coat

skirting board

1 After chopping out cut blocks at each side of the opening, fill the first course as shown in steps 4–6 of the photographed sequence. Complete the course with a cut piece of block if necessary.

2 Spread mortar on top of the first course and place the second course of blocks, laying the first block at the opposite side to the first block in the course below so that the vertical joints are staggered. Complete the course with another cut piece of block.

3 Add as many more courses of blockwork as you can. Fill the gap between the last course and the lintel with bricks or pieces of block cut to the required height.

4 Apply a base coat of plaster to both sides of the blockwork, filling the area to within about 3mm (⅛in) of the level of the surrounding plaster. Scratch its surface with the tip of a plasterer's trowel to provide a key for the finish coat. Allow it to harden for two hours.

5 Apply a finish coat of plaster over the base coat, overfilling the area slightly, then rule it off (see Points to Remember box) and polish the surface with a wet trowel.

6 Fit matching replacement skirting to both sides of the wall, using either nails or screws.

7 Redecorate when the plaster has fully dried.

opening up a disused fireplace ⚒

If your house has a fireplace that was blocked off the last time fireplaces went out of fashion, reinstating it is a relatively straightforward job. The amount of work involved depends on how the fireplace opening was blocked off, and on whether the old fireback was removed or left in position. Opening up a fireplace is very messy, so remember to roll back the carpet and put down a dustsheet before you begin work.

removing the infill

Tap the face of the chimney breast to discover if the infill is solid or hollow. Prize off the skirting board across the face of the chimney breast, saving it for later reinstatement at either side of your new fireplace. There should be an airbrick or ventilator in the face of the chimney breast to ventilate the flue. Start work by removing this – chop out a masonry brick from a solid infill, and unscrew a metal or plastic ventilator from a board infill. Then shine in a torch to see if the old fireback is still in place. If it is, you will only have some making good to do once the recess is reopened. If it is not, you will have to buy and fit a replacement, or get a builder to do the job for you.

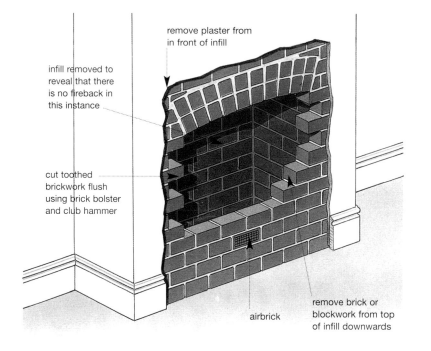

remove plaster from in front of infill

infill removed to reveal that there is no fireback in this instance

cut toothed brickwork flush using brick bolster and club hammer

airbrick

remove brick or blockwork from top of infill downwards

1 With masonry infill, chop off the plaster at skirting board level, working from the centre of the chimney breast outwards. This will reveal the edges of the infill.

2 Chop off plaster up the sides of the infill until you reach the top, where the infill will have been butted up against the lintel spanning the original fireplace opening.

3 If there was no airbrick to break out earlier, chop out an infill brick or block at one top corner of the infill. Chisel out its mortar joints after drilling a series of almost-overlapping holes into them with a drill and masonry drill bit, and lever it out.

4 Work across and down the infill, chipping out one brick or block at a time. If the fireback is still present, take care not to knock pieces of masonry inwards against it or you might crack it. Clear away the debris as you proceed.

5 If infill bricks have been toothed (bonded) into the masonry at each side of the opening, cut them flush with the original brickwork with a sharp downward blow using a brick bolster and club hammer.

REMOVING BOARD INFILL

With board infill, make a test drilling to see if plasterboard or a manufactured timber board has been used. If it is plasterboard and there was no air vent to remove, make a hole in the centre of the panel with a hammer and simply pull the pieces of plasterboard away. If plywood or another type of board was used, insert a padsaw or power jigsaw blade into the drilled hole and cut outwards towards the edges of the infill. Prize the cut sections of board away. Then look to see how the supporting framework of battens has been attached to the edges of the recess, and undo any screws you can locate. If there are no screws, assume that masonry nails have been used and lever the battens away carefully with a wrecking bar.

If the original fireback remains, you may have to patch it up in places. If it is missing, you will obviously have to buy a replacement one and install it. Start by measuring the width of the opening and order a new fireback to fit the space. Standard fireback widths are 400 and 450mm (16 and 18in),

but larger sizes are available if you want a more impressive fireplace. You will also need some lightweight mortar (made with vermiculite and lime), some corrugated cardboard, fireproof rope, brick rubble (use the infill you removed earlier) and fire cement.

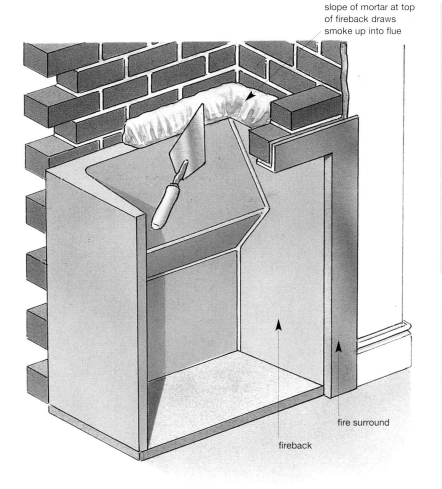

slope of mortar at top of fireback draws smoke up into flue

fire surround

fireback

PATCHING UP AN OLD FIREBACK

If an existing fireback is sound but cracked, you can repair it with fire cement. Let the fireback cool for a couple of days if it has had a real fire in it. Brush off soot with a wire brush and rake out the cracks with an old screwdriver, undercutting the edges. Wet the cracks to help the cement to stick, then fill them flush with fire cement. Smooth the filler with a wet paintbrush and allow it to harden for several days before relighting the fire.

safety advice

Wear safety gloves, goggles and a face mask when removing infill to prevent dust inhalation and injuries from flying debris. This is particularly important when working with masonry infill. It is also a good idea to enlist advice from a professional builder if you intend to have a real or fuel-effect fire installed, to ensure that the flue is sound and safe to use.

51

1 Separate the two halves of the fireback by tapping along the recessed cutting line with a club hammer and brick bolster.

2 Mix some mortar using four parts vermiculite (a lightweight granular insulation material) and lime, and place a bed of it where the base of the fireback will sit. Set the fireback in place, then pull it forward and trap lengths of fireproof rope between it and the edge of the fireplace opening.

3 Cut two strips of corrugated cardboard to match the height of the fireback and place them behind it, held against it with dabs of mortar. This will burn away when the fire is lit to leave an essential expansion gap behind the fireback. Then fill the space behind the fireback with mortar, bulked out with broken brick – you can use the remains of the masonry infill for this.

4 Bed some mortar on top of the lower half of the fireback and stand the top half in place on it. Neaten the joint and carry on filling behind the fireback.

5 Once the infill is level with the top of the fireback, add more mortar to form a slope up to the rear face of the flue. This forms a narrow throat that draws the smoke from the fire up into the flue.

6 Use fire cement to seal the edges of the new fireback to the fire surround, and to cover the fireproof rope.

blocking off a fireplace ⚒

Houses built before central heating became commonplace had a fireplace in every room. Today, even fireplace lovers are often happy with just a feature fireplace in the living room and may want to block off any remaining fireplaces in other rooms. However you decide to tackle this job, the one vital requirement is that the flue remains ventilated. Otherwise, condensation can form within it, soak into the chimney breast brickwork and eventually surface to ruin your decoration.

Before beginning any work, you have several decisions to make if you plan to decommission an old fireplace. The first concerns whether to strip out the old fireback and empty the fireplace recess, or whether to leave it in place. The former option is the better choice if you are certain the fireplace will never be used again, but it will make a lot of mess. The second decision concerns how to block off the opening. You can fill it with brick or blockwork, or panel it with plasterboard supported on a timber frame. Again, the former is the more professional option, the latter the quicker one.

tools for the job

safety goggles & dust mask

gloves

club hammer

brick bolster

wrecking bar

screwdriver

hawk

bricklaying trowel

plastering trowel

trimming knife

shovel

1 Use a brick bolster and club hammer to break the mortar bond between the raised hearth slab and the floor-level constructional hearth beneath it. Lever it up with a wrecking bar and get help to lift and remove it – it is too heavy to lift on your own. Then put down dustsheets in front of the fireplace opening.

2 Use a brick bolster and club hammer to chop away the plaster at the sides of the old fire surround to expose any fixing lugs. Undo them if you can with a screwdriver, with the aid of some penetrating oil to free rusty threads. Prize away the stubs of skirting board at either side of the surround, and save them as samples so that you can buy a matching length to cover the face of the new chimney breast surround when you have finished.

3 Insert the end of the wrecking bar between the surround and the chimney breast, first at one side and then at the other, and lever it away from the wall. Again, enlist help

to prevent it from toppling forwards and to lift and carry it away. Remove the grate if it is still in place.

4 Smash the old fireback with a brick bolster and club hammer. Wear safety goggles and a dust mask to protect yourself from dust and any flying debris. Lift out the sections of fireclay as you break them up, and put them straight into strong rubble sacks. Soak the fireproof rope around the perimeter of the opening with water to prevent asbestos fibres from getting into the air, cut it away with a trimming knife and put it in a plastic bag. Seal it, label it 'asbestos waste' and contact your local authority for advice on disposing of it safely.

✋ safety advice

Asbestos is a fibrous substance that can be woven with other materials to produce items that are highly heat resistant and have excellent insulation properties. However, it is a highly carcinogenic substance and the safety guidelines outlined in step 4 must be followed stringently. Never take risks when working with asbestos.

5 If the infill behind the fireback is a solid mass of mortar and broken bricks, break it up bit by bit with a brick bolster and club hammer. If it is loose rubble, simply shovel it out of the recess. Bag up all the rubble and remove it from the site to leave the fireplace recess empty. Use a vacuum cleaner to remove as much dust from the area as possible.

6 Use bricks or lightweight blocks to create a solid infill across the opening of the fireplace. Spread a line of mortar across the hearth and bed the first course in place. Cut the last brick or block to fit the space as

necessary, and use the offcut to start the next course so that the vertical joints will be staggered. (Refer to pages 48–9 for detailed information on blocking up an opening.)

7 Include a terracotta airbrick in one of the first few courses of bricks to ensure that the flue will be ventilated. Complete the infill, cutting bricks or blocks to size as necessary to fit the final course beneath the lintel that bridges the opening. Use mortar to fill any irregular gaps at the top of the space, and neaten the pointing.

8 Apply a base coat of plaster over the infill, recessing it by about 3mm (⅛in) to allow for the finish coat, and key it with a series of criss-cross strokes with the edge of a plastering trowel. Allow it to set hard, then trowel on the finish coat flush with the surrounding plaster. Polish it smooth with a wet trowel. Allow it to dry thoroughly before you redecorate it. Finally, cut and fit a new length of skirting board to the face of the chimney breast.

PANELLING THE OPENING

You may find it easier to fill the opening with plasterboard after stripping out the fireback rather than block or brickwork. If so, cut four pieces of 50 x 25mm (2 x 1in) softwood batten to fit the space. Glue and nail the top piece to the top ends of the two side pieces – you cannot nail it up into the lintel to fix it in place. Then secure the side pieces to the inner edge of the opening with masonry nails, and add the fourth batten across the hearth. Set the face of the battens back by 12mm (½in) so that the plasterboard plus a skim coat of plaster will end up flush with the surrounding plaster. Cut a hole in the plasterboard with a padsaw that is of the appropriate size to insert a plastic ventilator, then fix the plasterboard to the battens with galvanized plasterboard nails. Apply the finish plaster, then fit the ventilator when it has set hard.

tips of the trade

• **Capping the flue** – If the flue has an open pot at the top, it is a good idea to have it capped to stop rainwater from entering the flue. The simplest way of doing this is to fit a clay or metal hood top or flue vent into the top of the pot. Tackle this yourself only if you are happy working at height and can set up a ladder easily to reach the top of the chimney stack. Otherwise, call in a builder to fit it for you. Although the job itself is a straightforward one, it is not worth taking the risk of doing it yourself if you are inexperienced, and especially if you do not have the appropriate ladder and safety equipment for roof work.

• **Sweeping the flue** – If the fireplace was used regularly in the past, make sure that you remember to have it swept to remove all the soot that will have built up in the flue before you block up the fireplace opening. Doing this will minimize the risk of staining occurring on the face of the chimney breast in future if condensation forms in the flue.

building a masonry fireplace ↗↗ surround

A masonry fireplace surround can be built from your own design using brick or reconstituted stone walling blocks, or by purchasing a complete kit as shown here. Fireplace surround kits are available in a range of different designs and finishes, and come complete with specially shaped components to form the hearth, the top of the fireplace opening and the mantelpiece.

tools for the job

pencil

straight edge

bricklaying trowel

mallet

spirit level

tape measure

power drill

masonry drill bit

pointing trowel

paintbrush

1 Unpack the surround kit you have chosen, familiarize yourself with the various different pieces and check the kit supplier's recommendation concerning the right mortar mix to use. Begin assembling the kit by placing the hearth slabs. This kit has one wide centre section and two narrower side pieces. Lay the wide piece first. Mark its outline on the constructional hearth, then spread a generous bed of fairly soft mortar within the lines. Add another couple of trowels of mortar to support the centre of the slab.

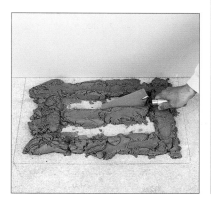

2 Set the centre section on its mortar bed and tamp it down with a mallet until the mortar joint is compressed to a thickness of about 25mm (1in). Trim away excess mortar all around, then check that it is perfectly horizontal by placing a spirit level on it, first across the width and then from front to back.

3 Lay the two side sections of the hearth in the same way, tamping them down until they are precisely level with the centre section. Trim away excess mortar and use a spirit level to check their alignment. Allow the mortar to set hard before proceeding with the next stage.

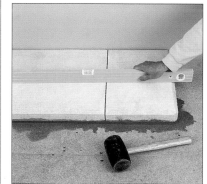

4 Stand the first upright section of the surround on the hearth and check the instructions supplied with the kit to find out where to position it. Note that the recessed section at the back should face the centre of the hearth. Mark the hole positions on the wall, then drill and plug the two screw holes. Repeat for the other section.

5 Move the first upright section back into position and fix it to the wall by driving two screws through the lug and into the wall plugs. Repeat for the second upright section at the other side of the hearth, and double-check that the gap between them is correct.

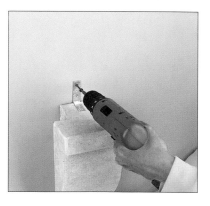

6 The two rectangular sections that frame the sides of the fireplace opening fit into the recesses in the main uprights. Slide each one into place, taking care not to knock the uprights away from the vertical.

7 Lift the crossbeam above the tops of the main uprights, align its ends with the recesses at each side and lower it carefully into place on top of the rectangular sections.

COMMISSIONING THE FIREPLACE

If you intend to light a real fire in your recommissioned fireplace, first you will need to seal the joint between the new surround and the existing fireback using fireproof rope and fire cement (both available from your kit supplier or local fireplace specialist). Cut the rope to the appropriate length and pack it in at the sides, then cover it with a layer of fire cement. Allow this to harden for a couple of days before lighting your first fire.

8 Complete the assembly by lifting the mantelshelf into place on top of the main uprights. It is the heaviest component of all, and may need two people to lift it. Check that it projects by the same amount at each side, then push it right back against the wall. Its weight will anchor it securely in place.

9 You may want to point the gaps between the hearth stones. Use the special mortar supplied with the kit, which dries to the same colour as the surround. Press it into the gaps with a pointing trowel, wipe away any excess and let it dry. You can fill the gaps between the sections that frame the fireplace opening if you wish.

If the fire surround is purely decorative, as here, paint the wall within the surround with black emulsion paint, ready for an electric fuel-effect fire to stand in front of it.

making changes outdoors

While your scope for executing masonry projects indoors is fairly limited, no such restrictions apply in the garden. Here, you can build walls to your heart's content – along boundaries, around a patio, as barriers to retain soil in terraces on a sloping site, or simply to hide a garden eyesore. You can add features such as a garden arch, and link different levels with flights of steps. You can build in brick or stone, creating formal or informal structures to suit the style of your garden. The two key points you need to remember are that every garden structure needs good foundations, and because you are building outside, everything you construct must be thoroughly weatherproof.

Combining different materials, in this case slate and stone block, can be used to create a pleasing contrast of colour and texture.

planning outdoor ↗↗ masonry projects

Whatever you plan to build in your garden, you need to do some planning and preparatory work first, even if you are just building a straight stretch of wall. This involves deciding what you want to achieve, where to site the various components of your scheme, what materials to use and how to organize the job into practical and achievable stages. You can then estimate and order materials with confidence, and tackle the job in an orderly fashion. You can also prepare the foundations that are essential for any garden structure.

making drawings

If you have a clear idea of what you want to create, buy some graph paper and pencils and start measuring things so that you can make a detailed scale drawing to work to. Begin with a plan so that you know where walls, steps and other features will be built. Add elevations to help you estimate materials accurately. Work to a sensible scale – 1:20 is suitable for most gardens –

and use metric measurements for accuracy. Make sure you know the sizes of the various materials you will be using, so that you can use your drawings for estimating quantities.

preparing the site

When you have decided what to build and where to site it, you must do some basic site preparation. Mark out the area of the work with string lines tied to pegs. Remove all plants,

grass, weeds and other vegetable matter, and dig away the topsoil, moving it to another part of the garden for redistribution. Cut back the roots of any large trees or shrubs well clear of the working area. As you dig, keep your eyes open for any buried services. Professionally laid cables and pipework should be buried at least 450mm (18in) down, but amateur supplies to garden buildings may have been laid at a far shallower depth.

laying foundations on a sloping site

The foundations should be laid as linked and overlapping steps. Each step should be a whole number of bricks in length, and the step height should be equal to a maximum of three courses of brickwork or one of blockwork (225mm/9in). The upper step should overlap the lower one by the length of two bricks or one block (about 450mm/18in). Peg a board across the trench at the step position to form the edge of the upper step and lay each concrete step in the same way that you would lay a concrete slab on a flat site (see opposite).

board pegged at step position to divide site into individual trenches for each step

concrete laid separately for each step by dividing site into trenches

upper steps overlap lower steps

For most garden walls, a concrete strip foundation is all that is required, but its size and positioning are important. It should be a minimum of 150mm (6in) thick, and up to 200mm (8in) thick on clay soils because of the tendency for clay to shrink and swell.

The strip should be placed in a trench 350 to 400mm (14 to 16in) deep, with its top one course of blockwork or three courses of brickwork below ground level. This allows soil to be placed right up to the foot of the wall, and helps protect the foundation from frost or accidental damage by digging close by. The width of the foundation strip should be twice the thickness of the wall you are building, up to a height of about 750mm (30in). For higher walls, increase the strip width to three times the wall thickness.

spirit level placed on straight edge laid over pegs to check they are level

trench for concrete foundation slab

pegs for gauging depth of concrete

concrete mix should be stiff enough to retain ridges formed in it with a shovel

1 Dig out the trench for the foundations to the required depth and width.

2 Hammer wooden pegs into the base of the trench to act as a guide for placing the concrete to the correct depth. Set them about 1m (3ft) apart and get them level by placing a straight edge and spirit level on adjacent pegs.

3 Estimate how much concrete you need for the foundation. For example, the strip for a wall 4m (13ft) long and 1.5m (5ft) high, built in 230mm (9in) thick brickwork on sound subsoil, should be 4.5m (15ft) long, 700mm (27in) wide and 150mm (6in) thick. Its volume is 4.5 x 0.7 x 0.15 cubic metres (15 x 2.25 x 0.5 cubic feet) – that is, 0.47cu m (16.87cu ft). To make this quantity, you will need about two-and-a-half 50kg (110lb) bags of cement and just under 900kg (2000lb) of all-in aggregate.

4 Mix the concrete (see pages 88–9 for more details). For in-ground foundations you should use a mix of 1 part cement, 2.5 parts sharp (concreting) sand and 3.5 parts of 20mm (¾in) aggregate (gravel). If you are buying all-in aggregate (mixed sand and gravel), mix 1 part cement to 5 parts aggregate. Use a bucket to measure the quantities accurately by volume, mix them thoroughly by hand or in a cement mixer, and add water until the mixture is stiff enough to retain ridges formed in it with a shovel.

5 Barrow the concrete to the trench, tip it in and compact it around the pegs. Use a piece of fence post or similar wood held lengthways in the trench to tamp it down, and add more concrete as necessary to get the surface level with the tops of the pegs. Do not worry about any slight irregularities in the surface. You can compensate for these when bedding the first course of masonry on the foundation.

6 Cover the concrete with polythene sheeting if rain or frost threatens, and allow it to harden for at least three days before starting to build on it.

FOUNDATIONS FOR STEPS

If you are building a flight of steps against a retaining wall, calculate its overall size. Then lay a concrete slab 100mm (4in) larger all around than the size of the flight and 100mm (4in) thick (150mm/6in thick on clay soils). See pages 88–91 for more details about laying concrete slabs. If you are building steps in a bank, only the bottom step needs a foundation strip. It should be 300mm (12in) wide, 150mm (6in) thick and 200mm (8in) longer than the width of the steps.

building a brick wall – 1 ⚒

For your first attempt at bricklaying, it is best not to be over-ambitious. Plan to build a straight section of wall no more than about fifteen bricks long and about eight bricks high, with the bricks laid end to end in stretcher bond. This will enable you to practise laying bricks level and in line, handling mortar, getting even joints and building to a true vertical. When you have mastered that, you can progress to forming corners and piers and building thicker walls.

tools for the job

string & pegs

hammer

bricklaying trowel

spirit level

brick bolster

club hammer

gauge rod (see box page 61)

bricklayer's pins

pointing trowel

1 The first step is to lay a foundation strip (see page 59). When you have done this, drive a peg into the ground just beyond each end of the strip and tie a length of string between the two to act as a guide for aligning the faces of the bricks in the first course of the wall. Position the string so that the bricks will sit centrally on the foundation.

2 Mix some mortar and trowel a strip about 150mm (6in) wide onto the foundation or patio. Level it roughly with the tip of a bricklaying trowel. Place just enough mortar to enable you to lay three or four bricks.

3 Working from one side of the foundation, set the first brick in place (some bricks have a recess, called a frog, which some people prefer to bed into the mortar, while others position it uppermost). Tamp the brick down into the mortar with the handle of the bricklaying trowel until it looks level. You can check this and adjust the level if necessary once you have laid several bricks.

4 Butter a generous amount of mortar onto the end of the next brick, pressing the mortar on firmly with the trowel held at 45° to the edges of the brick. If the mortar keeps falling off, try dipping the end of the brick in a bucket of water first.

5 Lay the second brick in line with the first, butting it up to the first brick so that the mortar on its end is compressed to a thickness of about 10mm (½in) between the two bricks. You may need to hold the first brick in place as you do this so that it cannot move. Tamp the second brick down level with its neighbour and check that the two bricks are in line. Continue in this way until you reach the end of the first course. Lay a spirit level along the top and against the faces of the bricks, and tap any that are high or projecting back into line. Then turn the spirit level through 90° and check that each brick is level across its width.

6 To ensure that the vertical joints are staggered in subsequent courses, you need to start the second course of the wall with a brick cut in half. To cut a brick, score a line across the flat side and place the brick on a soft surface (a bed of sand or the lawn will do). Align the edge of a brick bolster with the scored line and strike it firmly with a club hammer. The brick should cut cleanly with a single stroke.

7 Spread some mortar onto the first course of bricks and set the half brick in place to start the second course. Lay as many whole bricks as needed to complete the course, finishing it off with the other half brick that you cut in step 6. Trim off excess mortar from the joints as you work, then check line and level with a spirit level as before.

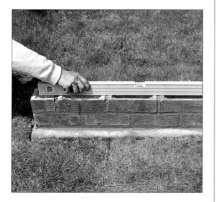

8 Add bricks at one end of the wall to start the third, fourth and subsequent courses, building upwards until the wall is eight bricks high. You will have to lay four bricks

in course three, three-and-a-half in course four and so on, up to one-and-a-half bricks in course eight. After laying each course, use a gauge rod to check that the horizontal (bed) joints are the same thickness (see step 10). The bricks now form even steps half a brick long running down from the end of the wall. This process is called racking back, and the object of the exercise is to build the wall by working from the ends towards the middle. Repeat the process at the other end of the wall.

9 Before proceeding any further, use a spirit level to check that the ends of the wall are truly vertical and that the face of the wall is flat. To check that the vertical joints are a standard width, rest the spirit level on the sloping brickwork. It should just touch the corner of every brick.

10 Hold the gauge rod against each end of the wall to check that the joint thicknesses are even all the way up. If not, knock down the faulty section and rebuild.

MAKING A GAUGE ROD

To check that mortar joints are the same thickness, make a tool called a gauge rod from a length of softwood. Mark lines on it to indicate the bricks or blocks and the mortar joints between them. In this project, for example, you should mark alternate brick widths (65mm/2½in thick) and joints (10mm/½in thick).

11 Push a bricklayer's pin into the mortar between the third and fourth courses at each end of the wall and tie a string line between them. Use this as a guide for laying the rest of the bricks in course three. Move it up a course at a time and complete another course of bricks, finishing with the eighth and final course laid with the frog down. You do not need a string line to lay this course. Once again, use a spirit level to check line and level, and make sure that all the vertical joints in the wall line up from course to course. You have just built your first wall.

building a brick wall – 1

61

building a brick wall – 2 ⤴

Once you have mastered the basics of handling bricks and mortar and building up courses, you can move on to turning corners and adding piers – essential supports if you are to build walls longer or higher than the one on the previous page. As with building a straight wall, the object of the exercise is to maintain the bond – the overlap of half a brick's length from course to course – to ensure that corners and piers do not form weak points in the wall structure.

tools for the job

string & pegs
hammer
bricklaying trowel
spirit level
builder's square (see box below)
brick bolster
club hammer
rubber mallet
gauge rod (see box page 61)
bricklayer's pins
pointing trowel

MAKING A BUILDER'S SQUARE

Cut three lengths of 50 x 25mm (2 x 1in) softwood – 400, 500 and 600mm (16, 20 and 24in) long. Glue and screw the two shorter pieces together at a right angle with a corner halving joint (formed by crossing the two lengths of wood and removing half the thickness of the wood from each piece to form a flush joint). Make a mark on the outside edge of the shorter length 300mm (12in) from the corner and on the longer one 400mm (16in) from the corner. Check that the distance between the marks is exactly 500mm (20in). Lay the 600mm (24in) length across the angle with its outer edge in line with the pencil marks. Glue and screw it to the other pieces, then cut off the overlaps at each end flush with the shorter pieces.

A simpler alternative is to cut a large triangle from the corner of a machine-cut sheet of plywood.

turning corners

If you build walls in stretcher bond, turning corners is straightforward. You simply place each corner brick at right angles to the one beneath it. This ties the two sections of the wall together and maintains the bond pattern in each course. The only cut bricks needed are in alternate courses at the open ends of the two sections, as for a straight wall.

1 Lay two foundation strips at right angles to each other (see page 59), then lay the first course of bricks forming one section of the wall (see page 60). Place the first brick of the second section at right angles to the corner brick of the first section, after buttering some mortar onto its end, and tamp it down level.

2 Place several more bricks in the first course of the second section of wall, and use a spirit level to check that they are truly horizontal and in line. Hold a builder's square in the internal angle between the two sections to check that they are at

right angles to each other, and adjust them if they are not. Complete the rest of the first course.

3 Start the second course by bedding two bricks in place at the corner, laid the opposite way around to the two in the first course. Tamp them down and get them level in both directions – along their length and across their width.

4 Build up the brickwork on both sides of the corner until the wall reaches its final height and you have just a single brick in the topmost course. This is the same process of racking back that you used in building a straight section of wall.

5 Hold a gauge rod against the corner to check that the joints are even, rebuilding the affected section if they are not (ideally you should check after each course). Build up the brickwork in the same way at each open end of the wall, and check the coursing there as well.

6 Place a spirit level or gauge rod on the sloping steps of brickwork to check that the racking back is even and the joints between the bricks are uniform – the level or rod should just touch the corner of each brick. Build up each section of wall as described in step 11 page 61.

building piers

Piers are also bonded into the wall structure for strength. In stretcher bond brickwork they can be one brick square and projecting from one face of the wall or, for maximum strength, one-and-a-half bricks square and centred on the wall. For walls built in stretcher-bond brickwork more than 450mm (18in) high, you need a one-brick pier at the end of the wall and at 3m (10ft) intervals along it. Position the larger piers in exposed locations. The maximum safe height for a stretcher bond wall with piers is 675mm (26in) – nine courses. Higher walls should be built 215mm (8½in) – one brick length – thick. This can be used up to 1.35m (4ft) without piers and up to 1.8m (6ft) with two-brick (440mm/17¼in square) piers.

projecting piers

To build piers one brick square at the end of a stretcher bond wall, place a brick alongside the last whole brick laid at the end of the first course. Place the first brick of the second course at right angles to them, then lay bricks in the second course as usual. Complete the pier with a half brick in this and every alternate course.

To build intermediate piers one brick square, place two bricks side by side at right angles to the wall face in the first course. To avoid the vertical joints aligning in the second course, centre a half brick over the two whole bricks in the first course, then lay a three-quarter brick at either side of the half brick. Complete the second course of the pier with a whole brick. Repeat this arrangement for alternate courses.

centred piers

You can build centred one-and-a-half brick piers in one of two ways. The first uses whole bricks for all courses of the pier, and the wall is tied to the piers with strips of expanded metal mesh bedded in the mortar every two or three courses. The second, stronger method bonds each course of the wall into the pier. This requires the use of half and three-quarter bricks in the pier structure to maintain the bond pattern.

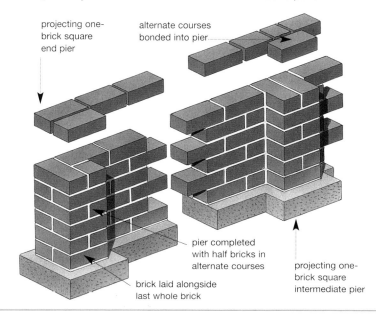

projecting one-brick square end pier

alternate courses bonded into pier

pier completed with half bricks in alternate courses

projecting one-brick square intermediate pier

brick laid alongside last whole brick

building a stone block wall ⚒⚒

Reconstituted stone walling blocks allow you to build natural-looking stone walls just as easily as if you were laying bricks. The blocks have faces and ends shaped to look like rough-hewn stone, but the top and bottom of the blocks are flat so that they can be laid in level mortar courses. Most ranges of blocks offer a choice of different lengths and heights, enabling you to build walls with the appearance of random stonework.

Visit suppliers such as garden centres and builders' merchants to select the blocks you want to use for the stone wall. There you will find the range of blocks available on display, giving you the opportunity to look at colour, texture and size options before deciding. The next stage is to plan the wall layout and estimate how many blocks are required. Accurate estimating is especially important if the wall is to comprise a mixture of block sizes. Design the wall on paper first and then count up how many blocks of each size will be needed to create the arrangement. Blocks tend to be sold in complete packs, but most suppliers will split packs if necessary. The smallest quantity they will supply is generally enough to build about 1sq m (10sq ft) of wall. In most cases suppliers will deliver direct.

tools for the job

bucket & mixing equipment
hawk
rubber mallet
bricklaying trowel
spirit level
gauge rod (see box page 61)
bricklayer's pins
builder's square (see box page 62)
pointing trowel

1 Lay the foundation strip (see page 59), then lay out the blocks dry. This enables you to check that the drawing translates correctly into three-dimensional reality, and provides an opportunity for correcting any

errors before actual construction begins. You can then pick up the blocks, course by course as you build up the wall.

2 Prepare a standard mortar mix, made from one part cement to five parts soft (building) sand. Then set up string guidelines as for building in brickwork, and trowel enough mortar onto the foundation slab to lay the first three or four blocks. Bed the first block in place on the mortar and tamp it down until it looks level. Butter mortar onto the end of the next block and lay it in the same way. Repeat for the next two blocks, then check with a spirit level that they are truly horizontal and aligned.

tips of the trade

Whatever the mix, there is a correct method for preparing mortar. Measure the proportions by volume, using separate buckets for cement and sand. Always mix the ingredients dry before adding water and make sure the mortar is not too sloppy or it will run and stain the faces of the stones.

3 Complete the first course by spreading more mortar on the foundation and laying further blocks. Start laying the second course, placing any double or triple-height blocks according to your sketch plan and checking that all the blocks overlap those in the first course to maintain the correct bonding pattern.

4 Place smaller blocks alongside the larger ones. There will inevitably be some alignment of the vertical joints where two or three smaller blocks butt up against a larger one. Use a gauge rod to check that the mortar joints between the stacked blocks are uniform in thickness. Add further mortar to the joints if required.

5 Continue building up the wall course by course, mixing large and small blocks according to your sketch plan. Make frequent checks with a spirit level as you build to ensure that the courses are truly horizontal and that the face and ends of the wall are both rising straight and not leaning or sloping too much. Complete the main part of the wall by adding the final course. The maximum recommended height for reconstituted stone block walls built with piers every 3m (10ft) is 625mm (2ft) in 100mm (4in) thick blockwork and 1.8m (6ft) in 210mm (8¼in) blockwork. It is advisable not to build any higher or the wall will become structurally unsound.

6 Now turn your attention to the pointing. Use a fairly dry mix for the pointing mortar in order to avoid staining the faces of the blocks. Take a sausage of mortar off the hawk with a pointing trowel and bed it well into the joint. Neaten the surface with the point of the trowel and leave to harden. Remove any droppings from

the faces of the blocks with a stiff brush when they have dried. Finish the wall by adding a layer of coping stones. These protect the top courses of the wall from rain and frost damage, and help to throw water clear of the faces of the blocks below. Set the coping stones in a generous mortar bed, tamping them down level with each other, then fill the joints with a further quantity of mortar.

BLOCKWORK WITHOUT MORTAR

Some manufacturers offer reconstituted stone blocks that are designed to be laid without mortar in low walls up to 625mm (2ft) high. The blocks have grooves on their undersides and ridges on their top surfaces, which interlock as the wall is built up. Other blocks are moulded to imitate drystone walling, with each manufactured block having the appearance of several interlocking stones. These can be laid with special thin-bed walling adhesive instead of mortar to give the appearance of a wall hand-built using individual stones. Matching coping stones are also available, moulded as stones set on edge so that you can finish the wall in the traditional way.

Coping stones add the finishing touch to a stone block wall and protect the blockwork from frost and damp. Simply tamp them down on a thick bed of mortar.

building a retaining wall ⚡⚡⚡

If you have a sloping garden or you want to create raised planters, you will need to build walls that hold soil behind them, so-called earth-retaining walls. These obviously have to be stronger than a free-standing wall because they act as a kind of dam, holding back not only the soil but also the considerable amount of moisture that it can contain after wet weather. This means building in 215mm (8½in) thick masonry, making provision for trapped groundwater to drain away.

When you are planning the design and siting of earth-retaining walls to terrace a sloping site, it is preferable to construct several shallow terraces rather than one or two tall ones (if you do build tall ones, they will need reinforcing; see box opposite). Smaller walls will be under less stress, and building steps to link the levels (see pages 72–5) will be much simpler. If you are creating several terraces, build the retaining wall that is furthest from the house first, so that you do not have to move materials from one terrace to the next. Earth-retaining planters are unlikely to be higher than about 900mm (3ft), and their box-like structure gives them more than enough strength to contain the soil that will be held within them, so reinforcement is unnecessary. Get professional advice before building earth-retaining walls higher than about 1.2m (4ft).

bonds for earth-retaining walls

It is possible to build a retaining wall with two parallel skins of stretcher-bond brickwork – known as double-thickness running bond – but the wall will not be very strong, even if the two leaves are tied together with cavity wall ties in the mortar beds. A stronger structure results if some of the bricks are laid end-on as headers, when they act as ties to hold the wall together. Two bonding arrangements commonly used for walls 215mm (8½in) – one brick length – thick are English and Flemish bonds.

In English bond the wall is built with different alternate courses, the first of stretchers laid side by side in running bond, the second of headers. At corners, a course of stretchers becomes a course of headers in the return wall, and vice versa. At ends and corners, a brick cut in half lengthways (a queen closer) is fitted before the last header to maintain the bonding arrangement. A variation called English garden wall bond has from three to five courses laid

as stretchers, followed by a single course of headers. It is sometimes used instead of pure English bond to reduce the amount of pointing required in the header courses, but is marginally less strong.

In Flemish bond, each course consists of a pair of stretchers followed by a single header, so that on the face of the wall each header is centred on the stretcher

below. Again, queen closers maintain the bond at ends and corners. The resulting wall is a little stronger than English bond because every course contains headers as through ties. A variation known as Flemish garden wall bond has a header after around three pairs of stretchers in each course, again to reduce the amount of pointing required.

ENGLISH BOND

alternate courses of parallel stretchers and headers

each course formed by pair of parallel stretchers followed by header

queen closer maintains bond at corners

FLEMISH BOND

queen closer maintains bond at corners

building an earth-retaining wall

Any earth-retaining wall needs solid, secure foundations. Excavate a trench to a depth of about 450mm (18in) and pack down a 150mm (6in) layer of hardcore using a fence post as a ram. Then drive in timber pegs to act as a depth guide for the foundation strip, and place a 150mm (6in) thick layer of concrete (1 part cement to 5 parts all-in aggregate) over the hardcore. Tamp it down with a beam held lengthways in the trench, level it and cover it with polythene. If you have clay soil, dig deeper so that you can put in a 300mm (12in) thick concrete strip. Allow the concrete to harden for at least three days.

tools for the job

spade
wheelbarrow
string & pegs
hammer
bricklaying trowel
spirit level
rubber mallet
brick bolster
club hammer
gauge rod (see box page 61)
bricklayer's pins
builder's square (see box page 62)
pointing trowel

1 You need to use special-quality frost-proof bricks for retaining walls. Reconstituted stone wall blocks are naturally frost-resistant. Lay the first course of brickwork, adjusting the thickness of the mortar bed to correct any unevenness in the foundation. Begin with a course of stretchers laid in running bond if you are building in English bond, as shown here, or pairs of stretchers followed by a single header if you are building in Flemish bond.

2 If the wall turns a corner, lay the first course of the return section of wall next. Place a queen closer behind the last stretcher at the face of the wall to maintain the bonding arrangement in the return wall. In English bond, the course continues along the return wall as a course of headers. In Flemish bond, place a pair of stretchers next to the queen closer, then a header and another pair of stretchers alternately along the rest of the course.

3 To allow water trapped behind the wall to drain away, either leave every other vertical joint open

in the bottom two courses of the wall to act as weepholes, or build short lengths of copper or plastic overflow pipe into the mortar bed above the first, second or third course of bricks – the first course is usually best.

4 Continue building the wall up to its required height in your chosen bond. Line the back of the wall face with heavy-duty polythene sheeting or give it two coats of liquid damp-proofing solution. This will prevent the masonry from becoming saturated, causing white efflorescence to appear on the exposed face of the wall – a problem that spoils the appearance of many retaining walls. Backfill behind the wall with gravel to a depth of at least 300mm (12in) to assist with drainage behind the foot of the wall. Allow the mortar to harden for a week before back-filling with soil.

REINFORCING WALLS

Walls higher than about 900mm (3ft) should be built around vertical steel reinforcing rods set into the foundation concrete. Build the wall in double-thickness running bond with the rods between the two leaves of the wall. Tie the two leaves together with cavity wall ties placed 450mm (18in) apart in every alternate bed joint, and in every course at wall ends and corners. Fill the space between the leaves with fine concrete.

building a screen block wall ✗✗

Pierced screen walling blocks allow you to build walls that act as a screen rather than a solid barrier. You can use them on their own or incorporate them as decorative panels in brick or block walls. Unique among building blocks, they are laid in vertical columns – an arrangement known as stack bonding – which means that a screen block wall is inherently weak unless it is reinforced during construction to compensate for the lack of bonding between the individual blocks.

Screen walling blocks are made to a standard size of 290mm (11½in) square, giving a modular unit 300mm (12in) square with a 10mm (½in) thick mortar joint. This means that any wall you build with them must be a multiple of 300mm (12in) in length and width. The maximum height for a wall built solely using screen walling blocks is 1.8m (6ft) – six courses. The blocks are usually 90mm (3½in) thick, and are white or off-white in colour.

Block manufacturers also make matching hollow pier blocks. These are 200mm (8in) high, so that three pier blocks coordinate with two walling blocks. They are made in four types – to form end piers, intermediate piers, corners and T-junctions – and, depending upon the type, they have one or more recessed faces into which the walling blocks fit. Any wall built with pier blocks must therefore have an even number of courses. Matching pier caps and 600mm (2ft) long wall coping stones complete the range.

tools for the job

bricklaying trowel
club hammer
rubber mallet
spirit level
pointing trowel

1 To build a free-standing wall, lay a foundation strip (see page 59), then spread a bed of mortar on it long enough to bed three blocks. Set the first pier block in place at one end

and tamp it down with the handle of a club hammer. Use a spirit level to check that it is standing squarely.

2 Butter some mortar onto one face of the first block and rest it on the mortar bed so that you can lower the mortared edge into the groove in the pier block. Tap it along gently until it engages, then tamp it down gently into the mortar bed. The blocks are fairly fragile, so try to tamp them close to the corners so that the force of the blow is transmitted down through solid material.

3 Butter the edge of the next block and bed it against the first, again tapping it horizontally until

the vertical joint is 10mm (½in) thick. Tamp it down level with its neighbour and trim off any excess mortar to prevent it from staining the block surfaces. Continue placing blocks in this way to complete the first course of the wall, finishing off with an end pier block. If the wall is longer than 3m (10ft) – 10 blocks – incorporate intermediate piers at a maximum of 3m (10ft) intervals.

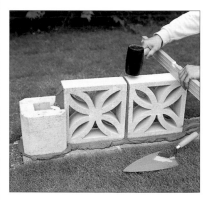

4 Spread some mortar on top of the first pier block at one end of the wall. Place the second pier block on it, tamp it down to a joint thickness of 10mm (½in) and check with a spirit level that it is level in both directions and that the pier is vertical.

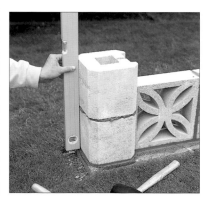

5 Repeat the process to place the third pier block. To reinforce the pier, fill its hollow central section with mortar, bedding it down inside the pier with a softwood offcut.

6 Place the second course of blocks on top of the first. Make sure that the first block engages fully in the groove in the pier blocks, and that subsequent blocks are perfectly aligned with the ones below. The top of the second course should be level with the top of the pier.

7 If you intend to build a wall higher than two courses, the pier blocks must be built around a steel reinforcing rod set in the foundations. Tie the wall blocks to the pier with a strip of expanded metal mesh bedded into the mortar in alternate courses and hooked over the reinforcing rod.

Add pier and pierced screen wall blocks as necessary to complete the wall. If the finished wall will be more than four courses high, add two more courses now and allow the

mortar to harden overnight before adding the next two.

Finish the wall by placing pier caps and coping stones on top of the wall. These could be matching or contrasting in colour, depending on taste. Check that the coping stones are level and that they overlap the wall blocks by an equal amount on each side of the wall. Mortar the joints between the coping stones.

tips of the trade

- **Matching mortar** – Since screen wall blocks are white or cream in colour, ordinary mortar will look much darker and will spoil the appearance of the wall. To get around the problem, order white Portland cement and the palest sand available from your walling supplier, and use these to make a matching mortar. If these materials are not available, the only alternative is to paint the wall with masonry paint or exterior-quality emulsion paint when the wall is complete.

- **Infill panels** – If you intend to use screen walling blocks as infill panels in solid masonry walls, remember that one block coordinates with four courses of brickwork. The blocks are a little narrower than a brick – centre the blocks on the brickwork if the wall will be viewed from both sides, but fit them flush with the face of the brickwork if only this face of the wall is visible.

Use a rubber mallet to tap the pier caps and coping stones into the mortar on top of the wall, and check with a spirit level to make sure that they are perfectly level.

building a brick arch ⟋⟋⟋

A brick arch is a striking way of framing a gateway or an opening in a high boundary wall (although building the latter is best left to the professionals). The structure is self-supporting once built, but you need the help of some timber formwork to support the arch bricks while the mortar sets. A two-ring arch is the best choice for a free-standing archway. A one-ring arch is weak and looks rather insubstantial, while a three-ring arch is too overpowering and looks better in a wall.

An arch works by transferring its weight downwards into the wall or piers that support it. If the arch is free-standing, the piers must be at least 215mm (8½in) – one brick – square, and it is generally best to err on the side of safety and build the piers measuring 330 x 215mm (13 x 8½in) with three bricks in each pier course. Each pier should have its own concrete foundation pad, 450mm (18in) square and 150mm (6in) thick.

tools for the job

bricklaying trowel

spirit level

tape measure & pencil

jigsaw

hammer

arch former (see step 2)

pointing trowel

raking tool or small cold chisel

1 Build the two piers, checking after each course that they are rising at the same level. Continue until you reach the springing point – the level at which the arch will begin. For

an archway 900mm (3ft) wide, piers consisting of 22 courses of brickwork will give adequate headroom.

2 Measure the width of the opening at the springing point and cut two semicircles of plywood to that diameter with a jigsaw. Nail them to wood offcuts to make an arch former measuring about 200mm (8in) from front to back. Set the former on props so that it sits level with the tops of the piers.

3 Place some mortar on top of each pier, next to the former, and set a brick in place on each one. Butt it up against the former and tamp it down in the mortar.

BUILDING A SEGMENTAL ARCH

You can use a variation on the brick arch technique to build a flatter arch, known as a segmental arch because its curve is a smaller segment of the circle's circumference than a semicircle. Use a pencil, some string and a drawing pin as improvised compasses to draw the curve you want on a sheet of board. Cut it out, check its appearance across the opening and use it to mark and cut a matching piece for the other side of the former. Remember that the flatter the arch, the more upright the bricks will be at either side of the opening. You will probably have to build the tops of the piers up level with the end bricks to improve their appearance.

4 Add bricks to each side of the former one by one, making sure that each one is butted against the former and that the wedge-shaped mortar joints between the bricks are the same width – around 10mm (½in) thick on the inside of the curve, and about 20mm (¾in) on the outside.

KEYSTONE ALTERNATIVES

Instead of using a brick as a keystone, you could use pieces of slate, plain clay roof tiles or quarry tiles. Simply fill the keystone position with mortar and push the pieces down into it one by one (you will probably need four or fives pieces).

5 If you have spaced your bricks carefully, there should be room for a single brick – the keystone – at the top of the arch. Butter mortar on both sides of it and slot it into place.

6 When the keystone is in place, hold a spirit level against the face of the arch to check that all the bricks are perfectly aligned.

7 Spread mortar on top of the first ring of bricks and build up the second ring in the same way, with wedge-shaped mortar joints. The second ring will have a slightly larger radius than the inner one, so the brick joints will not align after the first brick.

You will need more bricks for the second ring – typically 23 (compared with 19 for the inner ring) for an arch 900mm (3ft) wide. When you have added the second keystone, neaten all the visible mortar joints.

8 Leave the former in place for 48 hours to give the mortar time to set hard (if rain or frost threatens,

cover the top of the arch with polythene sheeting). Then remove the props carefully and let the former drop out without disturbing the brickwork. Point the joints on the underside of the arch to complete the job. Rake out the dried mortar to a depth of 4–5mm (⅕in) and replace it with fresh mortar using a pointing trowel.

When the arch is built, rake out the dried mortar from the joints and fill them with fresh mortar, either flush with the face of the bricks or recessed slightly, for a perfect finish.

building steps in a bank ⚡⚡

In a steeply sloping garden, the safest way of getting up and down sloping lawns is via a flight of steps, and the simplest way of building them is to use the bank to provide the step's foundations. As long as the subsoil is firm and has not been disturbed recently, careful cutting of the step shapes will provide a perfectly stable base for the flight. All you need is a single concrete foundation slab to anchor the lowest tread of the flight.

Garden steps should have treads at least 300mm (1ft) from front to back and 600mm (2ft) wide. Increase this to at least 1.2m (4ft) to create enough room for people to pass each other on the steps. The height of the risers will be governed by the bricks or blocks you use – two courses of brick or standard walling block topped by a paving slab will produce a step height of just under 200mm (8in). The treads should overhang the risers by about

25mm (1in). If the flight will be more than ten treads long, incorporate a wide landing halfway up the flight.

✋ safety advice

Treads and risers should be the same size throughout the flight and the treads laid with a slight slope from back to front so that rainwater drains off them and cannot freeze there.

👍 tips of the trade

To count the number of treads you will need, use a horizontal string line and a long garden cane stuck in the ground at the foot of the bank to measure the bank height. Divide this by the height of each tread to calculate how many steps you will need. Divide the length of the string line by the number of steps to check that the tread depth will be at least 300mm (12in).

preparing the site

1 Use pegs and string lines to mark out the sides of the flight and the positions of the tread nosings on the bank. Remember that treads need to be at least 300mm (12in) from front to back. Using paving slabs 450mm (18in) square gives a tread about 350mm (14in) deep, allowing for the depth of the riser that will be built off its rear edge.

2 Remove the transverse string lines and use a spade to cut out step shapes in the bank. The pegs will act as a guide to where the front of each tread should be. Work from the top of the slope downwards so that you do not break down the edges of the treads you have cut by standing on them.

3 Dig a trench at the foot of the flight and cast a 100mm (4in) thick concrete strip foundation about 300mm (12in) longer than the width of the flight and about 300mm (12in) wide (see page 59). Allow the foundation slab to set for three days before starting to build the flight of steps.

pegs and string act as guidelines for the slope and position of each step

dig trench at foot of flight of steps for concrete foundation slab

use a spade to cut out step shapes in bank

1 Spread a bed of mortar across the centre of the concrete foundation strip and bed the first course of bricks or blocks in it. Four bricks laid in stretcher bond make a flight of steps that is 900mm (36in) wide, ideal for paving with two 450mm (18in) square slabs per tread. Then cut a brick or block in half to start the second course of the riser. Bed it in place at one end of the riser and then complete the second course.

2 Infill behind the brick or blockwork of the first riser with crushed aggregate, which compacts better than hardcore because it contains a mixture of large and small pieces. Tamp it down well with a fence post or a similar sturdy, thick piece of timber, taking care not to disturb the bricks or blocks that you have just laid.

3 Spread a bed of mortar on top of the first riser and over the infill below where the edges of the paving slabs that are to form the treads will be. Bed the first slab in place, checking that it is level from side to side and has a slight slope towards its front edge to allow rainwater to drain off. Set the second slab of the tread alongside it, tamp it down level with its neighbour and trim off excess mortar from beneath the fronts of the treads.

4 Build the brick or blockwork of the second riser up at the rear of the first treads. Spread a mortar bed and set the bricks or blocks in it as before. Infill behind the riser and bed the next two treads in place, using the same method as before. Continue building risers and placing treads until you have completed the flight of steps. Finally, neaten the pointing and brush dry mortar into the gaps between the treads.

5 If the cut edges of the treads show signs of crumbling, build dwarf brick walls at each side of the treads to contain the soil.

lay bricks or blocks of first riser on bed of mortar above concrete foundation

use spirit level to check first tread is level

crushed aggregate infill behind riser

infill behind second riser and position second set of treads, then continue until required height is reached

build second riser at rear of first tread

building free-standing steps ⚒

If you are creating different levels in your garden through the use of retaining walls, you will need to build some steps to give you easy access from one level to the next. The easiest way to do this is to build a flight of free-standing steps between the two levels. This is more complex than building steps in a bank, since you have to create the entire structure yourself instead of letting the bank do the work of supporting the steps.

Plan the shape and dimensions of the new steps around the paving and walling you intend to use. It makes sense to match the bricks or blocks to those used for the existing retaining walls, and to choose paving that matches what you have elsewhere in the garden. Most paving ranges offer slabs in several different sizes, so you can choose whichever best suits the dimensions of your steps.

A free-standing flight of steps needs foundations to support the perimeter brickwork and the internal brickwork that supports the edges of the treads. Clear the site for the steps of vegetation and topsoil, and excavate to a depth of about 200mm (8in). Ram in a layer of hardcore about 100mm (4in) deep, and use sand to fill in the voids. Then place and level a 100mm (4in) thick concrete foundation slab about 100mm (4in) larger in each dimension than the flight of steps. Refer to pages 88–91 for more information on laying concrete slabs.

The flight of steps in the example illustrated here is built against a brick retaining wall four bricks high. It is four bricks wide, has risers two bricks high and treads formed of two 450mm (18in) square paving slabs. The top tread sits on the wall and the earth retained behind the wall.

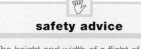

safety advice

The height and width of a flight of steps, as well as the materials used, contribute to their safety. Refer to pages 72–3 for advice.

1 Spread a bed of mortar along the line of the perimeter brickwork and lay the first course. Tamp them down with the handle of a trowel. Use a spirit level to check that they are level, and a builder's square to check the corners. Finish each course with a half brick.

2 Start the second course with a whole brick. On flights more than 450mm (18in) high, you need to tie the steps to the wall to prevent cracks from developing between the steps and the wall as time goes by. You can either chop out a half brick

tools for the job

pegs & string

garden spade

wheelbarrow

bricklaying trowel

spirit level

builder's square
(see box page 62)

brick bolster

club hammer

timber straight edge

in the wall and bond a whole brick from the steps into the recess, or use metal wall extension profiles (see pages 40–1).

3 Complete the second course of bricks. Once again, use a spirit level to check that the course is perfectly level, with all the bricks correctly aligned, and that the face of the brickwork is vertical.

4 Build transverse walls inside the brick box to support the second riser brickwork and the meeting edges of the slabs that will form the first tread. These can be of honeycomb construction (that is, the vertical joints do not need pointing).

5 Add two more courses of brickwork to form the second riser and the sides of the flight, and build up the rear internal support wall to the same height to support the meeting edges of the slabs that form the second tread. Use a spirit level to check that the wall is level.

6 Spread some mortar on top of the brickwork that forms the first step. Position the two slabs that will form the tread of the steps. Tamp them down level with each other, and with a slight fall towards the front of the step. Once again, use a spirit level to check the position.

7 Point the joints between the slabs and along their rear edges, and trim off excess pointing between them and the brickwork below. Place the slabs that form the second tread in the same way. If this tread is at the top of the flight, as in this example, rest the rear edges of the slabs on the wall behind.

POSITIONING HIGHER FLIGHTS OF STEPS

The picture sequence illustrated on these pages shows steps ascending at right angles to the face of the wall. If the retaining wall is more than six courses of brickwork in height, however, extra steps will be needed and the flight will have to extend farther away from the wall as a result of this. If this is not acceptable, either because you do not have sufficient space for them or simply for aesthetic reasons, turn the flight of steps through 90° so that the flight rises parallel to the wall face. Such a flight will then project onto the lower level by just the width of the treads.

Use the point of a bricklaying trowel to scrape away excess mortar from all the joints. The quality of the pointing can make a big difference to the finished look of the steps.

creating new outdoor surfaces

Most householders want to do more than just admire the garden from the house. They want to walk around it without getting their feet wet, sit outside without chairs and tables sinking into soft grass, and perhaps create decorative features with gravel or cobbles. They may even want areas of concrete to provide foundation slabs for garden buildings, or just to provide an inexpensive parking bay for the car. The most popular materials for creating patios, paths and other outdoor surfaces are paving slabs and interlocking paving blocks. Both come in a wide range of styles, shapes and colours and are very easy for the amateur landscape gardener to lay. All you have to decide is which to choose and where to lay them.

For areas with light traffic, such as patios or paths, slabs may be laid on sand rather than a mortar base.

laying slabs on sand ⌐

The simplest way of creating hard surfaces in your garden, such as patios and paths, is to lay paving slabs on a sand bed. Sand is easy for bedding and levelling and will provide a firm support for slabs as long as they are subjected only to pedestrian traffic. As the individual slabs are quite large, you can cover a sizeable area surprisingly quickly once the site is cleared and sand bed prepared. However, large slabs are heavy and anyone with back trouble should avoid laying them.

SLAB SIZES & FINISHES

Most paving slabs are square or rectangular, but sets are also available that build up to form circles. Square slabs range from 225 or 300mm (9 or 12in) up to 600mm (24in) in size, while rectangles start at 450 x 225mm (18 x 9in) and go up to 900 x 600mm (36 x 24in). This last is often used for laying pavements, but is perhaps too large for a small garden and certainly too heavy for one person to lift and lay. Finishes available include smooth, textured and riven – which is an imitation of split natural stone. Colours range from off-white and buff to red, granite and slate. Cheaper slabs are simply formed from cast concrete. These are relatively brittle and very difficult to cut accurately. More expensive slabs are hydraulically pressed, which makes them stronger and more easy to cut.

👍
tips of the trade

You need to plan the number of slabs you will need. To avoid unnecessary and time-consuming cutting, plan an area that is a whole number of slabs in length and width. Take rough measurements of the site, select the slabs you want to use and note the sizes in which they are manufactured. It is then a simple job to plan the layout in squares, rectangles or a mixture of the two, using graph paper to produce a scale drawing of the arrangement. Not only will this help when it comes to laying, it will also enable you to order precisely the right number of slabs in each size and so avoid unnecessary wastage.

tools for the job

tape measure
pegs & string
spade
hammer
wheelbarrow
shovel & rake
levelling board (home-made)
spirit level
mallet
brick bolster & club hammer
stiff-bristled broom

1 Mark out and clear the site, then excavate to a depth of 100mm (4in). Compact the subsoil if it has not been disturbed recently. If it has, dig a further 75mm (3in), then spread and compact a 75mm (3in) layer of gravel or crushed rock over the site. Tamp it down firmly with a length of fence post. Unless the paving will finish flush with an existing grassed area, edge the area with 100mm (4in) wide perimeter boards to retain the sand bed. Use sawn preservative-treated 25mm thick timber, nailed to timber pegs hammered into the subsoil.

2 Bring the sand to the site in a wheelbarrow and rake it out to a thickness of 50mm (2in). Use a piece of perimeter board as a levelling board to smooth over and level off the sand. Rest a spirit level on top of the board as you work to check that the sand bed has a slight fall away from the house for drainage purposes. Kneel on a board offcut as you work so that you do not disturb the sand.

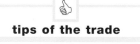
tips of the trade

To estimate how much sand to order, measure the length and width of the paved area in metres and multiply the two together to get the area in square metres. Then divide this by 20 to get the volume in cubic metres for a layer 50mm (2in) thick. Add 10 per cent to the total volume to allow for uneven subsoil and filling joints.

3 Place the first slab at one corner of the area and tamp it down into the sand bed using either a mallet or the handle end of a club hammer. Check that the slab is level in one direction and has the correct slope in the other.

4 Lay the next slab, either butting it up to the first one if you want close joints, or use wooden spacers if you prefer wider joints. Tamp it down and check that it is level with its neighbour, with the same fall away from the house. If the slab sits too low, lift it and add a little more sand underneath, then reposition.

5 Continue laying slabs along the first row, removing the spacers as soon as each slab is surrounded by further slabs on either side. Follow your plan unless you are laying only square slabs in a chequerboard pattern of rows and columns.

6 Once each row is complete, lay the levelling board along the row to check the slabs are lying level. Then check the fall down each row, and make any adjustments.

7 If your design is such that you need to cut a slab, mark the cutting line in pencil or chalk and use a brick bolster and club hammer to make a shallow cut across the slab. Then place the slab on the sand and strike the slab harder so that it snaps along the cut line. If you have a lot of cutting to do, it is a good idea to hire an angle grinder, which will make the cuts quickly and cleanly. When you have laid all the slabs, remove

the last spacers and shovel some sand onto the surface. Use a soft-bristled broom to brush the sand into all the joints between the slabs, then brush off the excess.

tips of the trade

If you are laying paving in an area that has a manhole, do not simply pave over it. The sand will enter the drains, and you will have an awkward job of lifting slabs to gain access. Get a builder to raise and reposition the frame so that the cover will be level with the new paving (or tackle the job yourself by adding a course of bricks around the top of the chamber).

There is no need to fill the joints with mortar when laying slabs on sand – it is enough simply to brush a further quantity of sand into the joints.

laying block pavers ⚒

Block pavers are the most popular choice for outdoor surfaces. Their advantages are that they are small and easier to handle than slabs, they are designed to be laid on a sand bed, and there is no pointing to be done as they are butted closely together. In addition, unlike other dry-laid paving materials, they will withstand the weight of a car. The only drawbacks are that it takes longer to cover an area with blocks than slabs, and the paved area must have edging to retain the sand bed.

Block pavers are generally rectangular and 50–60mm (2–2½in) thick. The most common size is 200 x 100mm (8 x 4in), but many other sizes are available. They also come in a wide colour range. The surface is slightly textured, with the block edges bevelled to emphasize their outline when laid. Either the blocks themselves or matching kerb blocks may be used as perimeter edging – these will need to be set in mortar.

Block pavers can be laid in various designs, from simple basket-weave to intricate herring-bone patterns. Most designs are laid square to the edge restraints, but you can also lay the blocks at an angle – generally 45° – running across the site. Note that the sand used for the bedding layer should be sharp (concreting) sand, not soft (building) sand, which does not compact as well and may also stain the blocks.

tools for the job

garden spade
tape measure
pegs & string
wheelbarrow & shovel
rake & tamping beam
levelling board
hired plate compactor (drives only)
spirit level
bricklaying trowel
rubber mallet
brick bolster
club hammer
hired block splitter (optional)
broom

1 Clear the site of vegetation and mark the area to be excavated. Dig out the area to a depth of about 100mm (4in) for patios and paths, and compact any disturbed subsoil – a thick fence post is used here. For a drive, excavate an additional 100mm (4in), then add a 100mm (4in) layer of hardcore or crushed rock. Compact this layer using a hired plate compactor (see box).

2 Position the edge restraints around the area to be paved. Set block or kerbstone edging in a mortar bed or, alternatively, nail preservative-treated timber edging to stout pegs. Allow mortar to harden for 24 hours before laying the blocks.

3 Fill a wheelbarrow with sand and tip the sand into the area to be paved. When all the sand is in place, rake it to a uniform depth of about 50mm (2in). Use a levelling board to smooth and level the sand across the area, working from a board so that you do not compact the sand bed by treading on it. Form stacks of blocks at intervals around the perimeter of the site.

LAYING BLOCK DRIVEWAYS

If you are laying block pavers as a driveway, you need to settle them into place to prevent them from subsiding under the weight of vehicles. To do this, hire a power tool called a plate compactor. This vibrates as you run it over the laid blocks, settling them into the sand bed and also compacting this so that it cannot itself subside. You should make one pass of the machine after placing the blocks, and another after brushing sand into the joints. This tool is also ideal for compacting any hardcore and crushed rock that is being used as a sub-base for any paving project.

4 Place the first blocks against one edge restraint, following whatever laying pattern you have chosen. If you are laying a patio or path, tamp them down into the sand bed using a rubber mallet. If you are creating a drive, simply set them in place. Make sure the blocks are butted closely together.

5 After laying about 1sq m (10 sq ft) or so of blocks, lay a spirit level on the levelling board and place it across the blocks in various directions to check that they are level and sitting flush with one another. Tamp down any that are proud of their neighbours.

6 Carry on laying blocks across the area, checking regularly that you are maintaining the pattern correctly. Depending on the pattern chosen and the shape of the site, you are likely to have to cut some blocks to finish the surface. Lay as many whole blocks as you can first of all, then insert the cut blocks, tamping each one into place.

7 You can cut block pavers with a brick bolster and club hammer, as shown here, but you will save time and effort (and spoil fewer blocks) if you hire a hydraulic block splitter for the day. Mark the cutting line in chalk, position the block in the cutter and pull down on the handle to split the block.

With all the cut blocks in position, lay a spirit level across them to check the levels once more and tamp down any that are proud of their neighbours. When you are satisfied, spread some fine, dry sand liberally over the surface and brush it this way and that until all the joints are filled. Sweep off the excess sand.

Make sure that you move the brush across the whole paved area, bending the bristles into all the joints to ensure that they are filled with sand for a perfect finish.

laying slabs on mortar ⁄⁄

If you want to use paving slabs for a drive or other area that will get more than just pedestrian traffic, you need to bed them on mortar over a concrete sub-base to ensure that the slabs have a continuous solid support and will not crack under the load. You can also lay slabs on mortar rather than sand for patios and paths. In this instance, a concrete base is not required but the subsoil beneath the slabs must be firm and well compacted.

Laying concrete from scratch in order to place slabs on top to create a driveway is an expensive way of creating this type of outdoor feature, but if the concrete is there already the job becomes a more economical proposition. As long as an existing concrete base is sound – in other words, not riddled with cracks and subsiding in places – and is at least 100mm (4in) thick, it will make the perfect foundation for a paved drive.

As with laying slabs on sand (see pages 78–9), the secret of success lies in careful planning. Take time choosing the slabs you want – you will have to live with the results for some time – and work out the layout to avoid having to cut slabs unless it cannot be avoided. On an existing concrete base you may have to mix slab sizes to ensure that you have whole slabs along all open edges of the base. Increasing or decreasing the spacing between the slabs may also help you avoid having to cut slabs.

tools for the job

tape measure

shovel

bricklaying trowel

rubber mallet

spirit level

timber straight edge

watering can

pointing trowel

stiff broom

brick bolster

safety goggles

work gloves

patios & paths

After planning (and if necessary sketching out a scale drawing) the slab layout, clear and mark out the site. If laying onto subsoil, make sure it is well compacted. Mix a fairly sloppy mortar mix in the proportions 1 part cement to 6 parts soft (building) sand. Compact the subsoil thoroughly and place five pats of mortar in the place where the first slab will be positioned, one under each corner and one in the centre. Lay the slab in place on the mortar and tamp it down with the handle of a club hammer or a rubber mallet so that it is perfectly level. Then repeat the process to lay other slabs across the area, checking the level from time to time with a spirit level resting on a timber straight edge. When you have laid all the slabs, brush a mixture of cement and sand into the joints and sprinkle water onto the paving with a watering can to dampen the mortar and make it set hard between the slabs. Once dry, the mortar locks the slab in place and ensures that it does not move.

driveways

1 If you are laying a driveway on a concrete base, brush the surface of the concrete with a stiff broom to remove any loose material, and treat it with a fungicidal wash to kill any weeds, lichen and algal growth. This should prevent any weeds or other plant growth from emerging through the joints in the paving for some time, though you will have to take measures to control this problem in the future. Spread a generous square of fairly sloppy mortar where the first slab will be positioned, and add two lines of mortar at right angles across the centre of the square – this is called box-and-cross bedding in the trade.

2 Place the first slab on the mortar and tamp it down well with a club hammer. This will compress the bedding into a continuous layer of mortar and bond the slab securely to the concrete. Use a spirit level to check that the slab is level (assuming that the concrete base beneath it is level, of course).

3 Place more box-and-cross mortar for the next slab, position it and put two 10mm (½in) wood spacers between it and the first slab. This ensures that the pointing gap remains constant across the whole paved area. Then tamp the slab down and check that it is level and aligned with the first slab.

4 Lay the rest of the slabs in the same way with spacers between them, working your way across the area row by row. You can remove the spacers as soon as each slab is surrounded on all sides by other slabs.

5 Allow the mortar to harden overnight. Mix a fairly dry mortar mix so that it does not stain the slabs, and work it into the joints with a pointing trowel. Use an offcut of wood to pack the mortar down well. Allow mortar droppings to dry on the slabs, then brush them off.

cutting slabs

1 If you have to cut just a few slabs, do it by hand. Wearing safety goggles and gloves, mark the cutting line and cut a shallow groove along it with a brick bolster and club hammer. Alternatively, score the line with repeated passes of the corner of the brick bolster.

2 Place the scored slab on a bed of sand and position the brick bolster in the centre of the line. Strike it firmly with the club hammer and the slab should split along the scored line. If you have to cut a lot of slabs, consider hiring an angle grinder for the day.

ENSURING EFFICIENT SURFACE DRAINAGE

With mortar-pointed slabs, rainwater will not drain away between the slabs in the way that it does when slabs are bedded on and pointed with sand. You should therefore build the slabs on a slope in order to prevent puddles from collecting and standing on the surface after it has rained. The slope should be of a continuous uniform gradient across the paved area.
If the paved area is next to the house, lay the slabs so that the slope falls away from the house. Aim for a slope of about 1 in 40 – that is, a 25mm (1in) drop for every 1m (3ft) of patio or drive width.

tips of the trade

- **Seal the concrete** – If the existing concrete slab is dusty, treat it with a solution of pva (polyvinyl acetate) building adhesive, diluted as recommended by the manufacturer for use as a sealer. Not only will this bond the concrete surface together, but it will also help the mortar to bond well to the concrete, ensuring that the slabs do not work loose as time goes by.

- **Plant the gaps** – If you want your paving to have an informal look, lay the slabs with wider than usual joints and fill them with soil rather than mortar to encourage grass, moss and low-growing plants. Using slabs in a mixture of different sizes also looks less formal than same-size slabs laid in carefully regimented rows.

laying crazy paving ⚒⚒⚒

Crazy paving is a paved surface created by bedding irregularly shaped pieces of stone in mortar and then pointing the gaps between the pieces. The stone is usually broken paving slabs, although natural stone can also be laid as crazy paving in the same way. It may have got its name from the crazing that affects the glaze on old pottery, as the appearance of well-laid crazy paving is similar. The stones have to be well fitted and the pointing carefully detailed to avoid a piecemeal finish.

The main attraction of crazy paving, apart from its rustic and informal look, is that the paving is cheap – broken slabs are widely available in large quantities from demolition contractors and also some local authorities. As a rough guide, one tonne of broken slabs will cover an area of 9–10sq m (100sq ft), but this depends on the thickness of the slabs. Ask your supplier for advice when ordering. You may be offered a choice of stones in one predominant colour and texture, or as a random mixture. Generally speaking, crazy paving looks better if all the pieces are more or less the same shade.

When the delivery arrives, sort the stones into groups. Pieces with two square adjacent edges will form corners, while those with one straight edge will serve as perimeter stones. Separate what is left into large irregular pieces for the centre of the paved area, and small infill pieces. Stack the groups around the paving site so that you have got them to hand while you work.

The best base for crazy paving is an old concrete slab in need of a facelift. The alternative is a layer of crushed rock or hardcore, laid over well-rammed subsoil. Brush old concrete to remove loose bits from the surface, treat it with a fungicide if it is covered in lichen and green algae, and seal its surface with diluted pva (polyvinyl acetate) building adhesive if it is very dusty. Clear a virgin site of vegetation, skim off the topsoil and bed down a 75mm (3in) layer of crushed rock, ready for the mortar bed that will take the paving.

tools for the job

shovel

wheelbarrow

hired cement mixer

bricklaying trowel

rubber mallet

spirit level

timber straight edge

club hammer

brick bolster

pointing trowel

spot board for mortar

stiff broom

1 Mix your first batch of mortar using a hired cement mixer – you will need so much mortar for bedding and pointing that hand mixing is not really an option for this job. Use a mix of 1 part cement to 5 parts sharp (concreting) sand, and make it relatively sloppy. Start work at one corner of the site, shovelling enough mortar onto the area you are paving to cover about 1sq m (10 sq ft) at a time. Trowel it out to a thickness of about 50mm (2in).

2 Select a corner stone and tamp it down into the mortar bed using a mallet or the handle of a club hammer. Use a spirit level to get it level, and make sure it is aligned with the edges of the slab if you are laying on concrete. Otherwise, use string lines to help you keep the edge stones accurately in line.

3 Choose a large perimeter stone that fits well against the corner stone – aim for a pointing gap of no more than 25mm (1in) in width, or your crazy paving will look very gappy. Set it in place and tamp it down, using a mallet or the handle of the club hammer, then check that it is level and in line with the corner stone.

4 Continue laying perimeter and corner stones all the way around the site until the edge is complete. As you bed the stones, mortar will squeeze up between them. Leave it, as it will reduce the amount of pointing needed later, but make sure it does not sit proud of the surface of the stones.

5 Shovel in more mortar, spread it out and start bedding large irregular stones in the centre of the area, working from one end to the other. Use a brick bolster and club hammer to trim stones to improve their fit. Check that each stone is level as you tamp it down by placing a timber straight edge across the paving with its ends on the perimeter stones. Kneel on a board as you work across the site to avoid disturbing the pieces you have already laid.

6 Use smaller pieces of stone to fill in the remaining gaps, again trimming them to size first if necessary. Use any remaining mixed mortar to fill the gaps between the

stones to within about 5mm (¼in) of the surface of the paving. Allow the mortar to harden overnight before starting on the pointing – the most time-consuming part of the job.

7 Make up a pointing mortar of 1 part cement to 5 parts soft (building) sand. It should be somewhat drier than the bedding mortar to avoid undue staining of the stone surfaces. Trowel it into the joints from a spot board, then draw the side of the trowel blade along the edge of each stone to recess the mortar about 3mm (⅛in) below it and leave two sloping bevels meeting at a central ridge.

8 Allow any loose mortar that you have dropped to dry on the stones – lifting this when it is wet will leave stains. Once the pointing mortar has set hard, brush the paving surface to remove the debris.

Crazy paving has a rustic, informal look that is perfectly enhanced by a busy planting scheme, with flowers and shrubs allowed to overlap onto the paving.

laying gravel & cobbles ↗

A gravel path or drive makes an attractive contrast to flat paving materials, especially if the aggregate is chosen with care. Strictly speaking, gravel is made up of small water-rounded pebbles and is available in a range of natural earth shades. Crushed stone aggregates have rough-edged stones and are available in colours ranging from white and grey through reds and greens to black. These can be laid as a single colour or can be mixed if preferred.

tools for the job

tape measure

spade

handsaw

club hammer

claw hammer

bricklaying trowel

shovel & wheelbarrow

garden roller or tamping beam

rake

timber straight edge

using cobbles

Cobbles are large rounded river stones. They make an uncomfortable surface to walk on but can look very attractive when used in the garden to provide a visual counterpoint to flat surfaces – perhaps as a border to a path. They can be loose-laid, but are better bedded in mortar.

1 To create a cobble feature, complete the surface that it will complement – a brick or block border, for example. Then spread a bed of

fairly sloppy mortar about 50mm (2in) deep and start placing individual cobbles in it to about half their depth.

2 Use a timber offcut and a club hammer to tamp them down into the mortar bed until they are reasonably level. This will ensure that they do not work loose as time goes by. When the mortar has dried, brush away any excess and apply a coat of clear silicone masonry sealant to the stones to give them a permanent wet look and enhance their colours.

using gravel

Gravel surfaces are satisfyingly crunchy to walk on or to drive over, and are an excellent burglar deterrent – you cannot cross gravel quietly, even on tiptoe. However, they do have several practical drawbacks. For a start, they need some form of edge restraint, such as pegged boards or concrete kerbstones, to stop the gravel from migrating onto lawns or into flowerbeds. You also have to rake them regularly to keep them looking neat and tidy, and they

will need treating with weedkiller from time to time. Another drawback is that they can attract local cats and dogs, which regard gravel as an ideal outdoor litter tray. Last of all, pushing a laden wheelbarrow across gravel is like wading through quicksand.

If you decide to create a gravel feature in your garden, first choose which type of aggregate to use. Then measure the area you intend to cover and the depth to which you want to fill it so that you can calculate the volume you require. A depth of 50mm (2in) is adequate for a path, but 75mm (3in) is better for a drive.

You will need a bulk delivery for all but the smallest projects. A cubic metre (35cu ft) of gravel or aggregate will cover about 20sq m (215sq ft) to a depth of 50mm (2in), and 13sq m (140sq ft) to a depth of 75mm (3in). If your supplier delivers by weight, one tonne will cover about 7.5sq m (80sq ft). Bulk loads are delivered loose or in large canvas slings – these are preferable to a big heap because the gravel is contained better while you move it to its final destination. For small projects, you can buy gravel and other decorative aggregates in 25, 40 and 50kg (45, 90 and 110lb) bags.

1 Measure out the site and clear vegetation and topsoil. Compact the subsoil thoroughly. Put the edge restraints in place, either nailing preservative-treated timber to stout pegs or bedding kerbstones in a fine concrete mix. If you are using boards, secure them with pegs every 1m (3ft) or so to prevent them from bowing.

tamping beam such as a length of fence post to pack down the layer of rock. To test whether you have tamped it down sufficiently, walk on the surface – if you do not leave footprints, then the rock has been tamped down enough.

shovelful up to the level of the edge restraints. Take care not to disturb the compacted base layer. Rake the gravel out level, then place a timber straight edge on the edge restraints and draw it across the site to identify any high or low spots. Rake off the former and fill the latter with more gravel. Once the gravel is level, roll or tamp it to bed it down well and leave the finished surface about 25mm (1in) below the tops of the edge restraints.

2 Lay a proprietary weedproof membrane over the site to discourage deep-rooted weeds from growing through the gravel. Overlap strips of membrane by at least 100mm (4in), and trim the edges.

3 You can lay the gravel directly over the membrane, but if the subsoil is soft it is best to put down a layer of crushed rock or fine hardcore first. Shovel it onto the membrane to a depth of at least 50mm (2in).

4 Compact the crushed rock with a heavy garden roller if you have one. Otherwise, use a heavy timber

5 Use a wheelbarrow to transport the gravel to the site and tip it out in heaps. When you have done so, spread out the gravel shovelful by

It is important to rake the gravel level in order to achieve an attractive finish, and it is a good idea to do this regularly to keep your gravel feature looking as good as new.

laying concrete – 1 ⁄⁄

Concrete can be used to create a driveway, a patio or a path but suffers by comparison with paving slabs and blocks as far as appearance goes. It can also form bases for lightweight garden buildings such as summerhouses, sheds and workshops, and is the material used to create foundations for all sorts of garden structures such as walls, arches and steps. Its main advantage over other materials used for creating outdoor surfaces is that it is very economical.

ingredients

Concrete is a mixture of coarse and fine aggregates – stones up to around 20mm (¾in) in diameter with smaller stones and coarse sand – that is bound together into a solid matrix by cement. You can buy the ingredients separately from builders' suppliers and mix them yourself, buy dry ready-mixed bags of cement and aggregate (ideal for small jobs) or order ready-mixed concrete (best for large areas).

Ready-mixed concrete may be delivered by a large truck mixer with its familiar slowly turning drum, or by a smaller vehicle that carries dry cement, aggregates plus a cement mixer and can mix the amount you need on the spot. Truck mixers can deliver up to about 6cu m (200cu ft) of concrete from their chutes directly to the site. Smaller vehicles mix by the barrowload, which you then have to move from truck to site.

The ingredients of a concrete mix depend on the use to which the material will be put. The three standard formulae are given in the table below, along with the quantities you need to make 1cu m (35cu ft) of concrete. All-in aggregate is a mixture of sharp sand and 20mm (¾in) aggregate. Always mix ingredients by volume, using separate buckets or similar containers of the same size for cement and aggregate. Mix batches based on 1 bucket of cement plus the relevant numbers of buckets of sand and aggregate.

USE	MIX	PROPORTION	AMOUNT PER CU M (35.3CU FT)
General-purpose cement (most uses except foundations and exposed paving)	cement	1	6.4 bags (320kg/700lb)
	sharp sand	2	680kg/1500lb (0.45cu m/15.8cu ft)
	20mm (¾in) aggregate	3	1175kg/2600lb (0.67cu m/23.6cu ft)
	OR all-in aggregate	4	1855kg/4100lb (0.98cu m/34.6cu ft)
Foundations (strips, slabs and bases for paving)	cement	1	5.6 bags (280kg/615lb)
	sharp sand	2.5	720kg/1600lb (0.5cu m/17.6cu ft)
	20mm (¾in) aggregate	3.5	1165kg/2560lb (0.67cu m/23.6cu ft)
	OR all-in aggregate	5	1885kg/4150lb (1cu m/35.3cu ft)
Paving (exposed slabs, especially drives)	cement	1	8 bags (400kg/880lb)
	sharp sand	1.5	600kg/1320lb (0.42cu m/14.8cu ft)
	20mm (¾in) aggregate	2.5	1200kg/2640lb (0.7cu m/24.7cu ft)
	OR all-in aggregate	3.5	1800kg/3960lb (0.95cu m/33.5cu ft)

mixing your own concrete

For hand mixing you will need a hard, flat surface. A sheet of exterior-grade plywood is ideal for protecting drives or patios, which should not be used unprotected because the concrete will stain them, however promptly you hose them down. Plastic trays 1m (3ft) across may be bought for mixing small quantities.

tools for the job

shovel

clean bucket

wheelbarrow

hired cement mixer

by hand

1 Measure out the sand and aggregate into a compact heap. Form a crater in the centre with a shovel and add the cement. Mix the ingredients dry until the pile is uniform in colour and texture. If you are using dry ready-mixed concrete, tip out the sack and mix thoroughly.

2 Form a crater in the centre of the heap and add water. The aggregate will contain a certain amount of water already, so the amount you need to add will be trial and error to begin with. After two or three batches, you will be better able to gauge how much to add.

3 Turn dry material from the edge of the heap into the central crater. Keep on mixing and adding a little more water in turn until the mix reaches the right consistency – it should retain ridges formed in it with

the shovel. If it is too sloppy, add dry ingredients, correctly proportioned as before, to stiffen it up again.

with a mixer

If you are using a cement mixer, set it up on its stand and check that it is secure. Put some aggregate and water in the drum and start it turning. Add most of the cement and sand, then water and solid material alternately, to ensure thorough mixing. Run the mixer for two minutes once all the ingredients are in, then tip out some of the contents into a wheelbarrow. The mix should fall cleanly off the mixer blades.

preparing formwork

Concrete foundations for walls can generally be poured straight into a prepared trench, since they will be hidden once the wall is built. However, surfaces such as paths, patios and bases for buildings need to have straight vertical edges, and the way to provide these is to lay concrete within what is known as formwork or shuttering – lengths of timber supported by stout pegs to form a mould for the concrete. The top edges of the formwork provide a levelling guide for the concrete, while its inner faces give the finished slab a neat moulded edge.

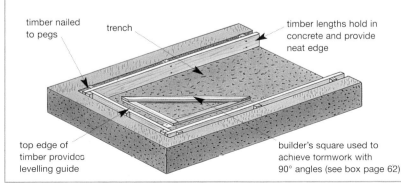

timber nailed to pegs

trench

timber lengths hold in concrete and provide neat edge

top edge of timber provides levelling guide

builder's square used to achieve formwork with 90° angles (see box page 62)

laying concrete – 2 ⚒

With the concrete mixed and the formwork in place, you are ready to start laying your concrete feature. Make sure that you have excavated the site to the correct depth and that you have placed and compacted a layer of crushed rock or hardcore if this is required – see pages 18–19 for details. Check that the formwork is square and level, or has a slight fall if drainage of rainwater is needed away from the house.

Large areas of concrete cannot be laid as continuous slabs or they will crack due to expansion and contraction. You must therefore divide the work into bays, each separated from its neighbour by an expansion joint of hardboard or similar material, if the concrete is being laid as a continuous operation. Where the area being concreted has a curved edge, make sure that expansion joints meet the curves at right angles. If you are using the alternate bay technique (see opposite page), an expansion joint is not necessary – simply remove the formwork between filled and unfilled bays before concreting the unfilled bays. The recommended maximum size of each bay is around 4 x 4m (13 x 13ft). On paths less than 2m (6ft 6in) wide, incorporate an expansion joint every 2m (6ft 6in).

tips of the trade

It is best not to lay concrete if frost is forecast because permanent damage will be caused if the water in fresh concrete freezes. If an unexpected frost sneaks up on you, lay polythene sheeting over the concrete and cover it with a layer of earth or sand. Leave this in place until a thaw sets in. Never lay concrete on frozen ground.

compacting the concrete

The most important part of concreting is compaction. For a concrete slab, the ideal compacting tool is a length of 100 x 50mm (4 x 2in) sawn timber used on edge, long enough to span the formwork. If you wish, add angled timber handles to each end of the tamper so that you and a helper can operate it in tandem.

Shovel concrete into the formwork, then rake it out level. Make sure that corners and edges are well filled. Add more concrete until the level is about 12mm (½in) above the top edges of the formwork.

Set the tamper in place across the formwork at one end of the slab and start tamping the concrete down. Use the beam with an up-and-down motion until the concrete is level with the top edge of the formwork, moving it along by half the beam's width after each tamping stroke. Do not simply scrape the tamper along the formwork. Continue tamping in this way until you reach the far end of the concrete slab, then repeat the process using a side-to-side sawing action to remove excess concrete. If the surface of the concrete still contains small voids, spread a thin layer of extra concrete and use the tamping beam again.

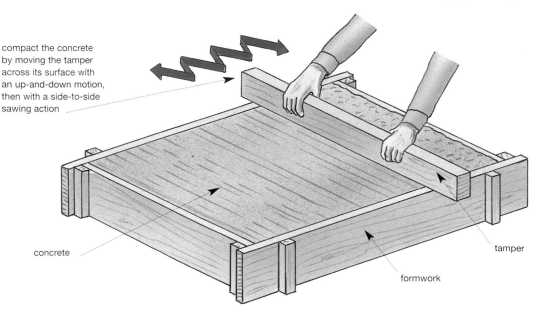

compact the concrete by moving the tamper across its surface with an up-and-down motion, then with a side-to-side sawing action

concrete

tamper

formwork

Cut strips of hardboard to match the thickness of the concrete and set them on edge with dabs of concrete against their sides. The top of the strip should be level and just below the top edges of the formwork. Then lay concrete up to the strip, working from both sides at the same time so that placing and compaction do not push the strip out of position. Incorporate similar joints where concrete slabs meet buildings. If there is an inspection chamber within the area being concreted, first place formwork around the chamber so that you can place a box of concrete around it. When it has set, remove the formwork and fit expansion strips all around it before concreting the rest of the slab. For concreting irregular-shaped areas, set formwork boards across the area to divide it into bays. These boards should always be at right angles to the perimeter boards.

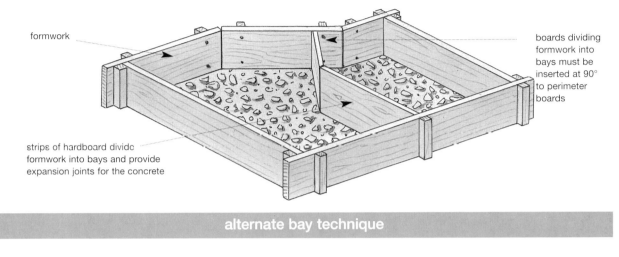

formwork

boards dividing formwork into bays must be inserted at 90° to perimeter boards

strips of hardboard divide formwork into bays and provide expansion joints for the concrete

alternate bay technique

If a slab adjoins a wall, you will not be able to use the tamper at right angles to the wall. Instead, you will have to lay alternate bays along the wall. Set up formwork as usual, with timber stop-ends dividing the bays. Place an expansion joint next to the wall and fill alternate bays, tamping the concrete parallel to the wall. Remove the stop-ends after 48 hours and concrete the remaining bays, using the hardened concrete edges of the first set of bays to guide the tamping beam. There is no need to fit expansion joints between the bays.

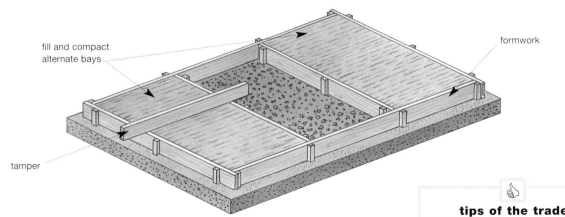

fill and compact alternate bays

formwork

tamper

FINISHES FOR CONCRETE

You can leave concrete with the rippled finish created by tamping but this is not very attractive. You can create a finer rippled finish if you do the final tamping by working the beam backwards and forwards with a sawing motion at right angles to the formwork. Brushing the tamped surface with a soft broom will flatten the ridges and leave a fairly smooth finish. A wood plasterer's trowel will give a fine sandpaper texture, while a steel float will create a fine, flat finish.

tips of the trade

Fresh concrete will weaken and crack if it dries too quickly, and this is a particular risk with thin slabs of large surface area. As soon as the concrete slab is laid and finished, cover it with polythene sheeting weighted down all around with bricks or baulks of timber to stop wind from blowing under it. Leave the sheeting in place for about three days.

plastering & lining

Before walls and ceilings can be decorated with paint, wallpaper or other finishes, they have to be given a smooth, flat surface. On brickwork and blockwork walls, plaster does the job. This is a powder based on gypsum that is made into a plastic mix with water and applied to wall and ceiling surfaces, where it dries to a hard coating. On timber-framed walls, a rigid sheet material called plasterboard is used to clad both sides of the wall structure. Plasterboard is also used to form ceilings beneath timber floor joists, and to line the inside of exterior walls as an alternative to plastering. There are several other plaster-based products that are used around the house for their decorative effect, including coving, ceiling centres and panel mouldings. This chapter tells you how to use them.

Ornate coving complements the elegant grandeur of this room, but plain coving would suit a more modern decor.

plastering masonry ⁄⁄⁄

The type of plaster most widely used for masonry is a mix based on a mineral called gypsum. This is usually applied as a two-coat system, with a thick undercoat applied first and thinner finish coat on top. Different types of plaster are used for undercoats and finish coats, and there are different undercoats for different backgrounds such as brick, aggregate blocks and thermal blocks. Ask your supplier for advice to ensure that you select the correct plaster for the job.

The job of the plaster undercoat is to smooth out any irregularities in the wall surface and even out differences in the rate at which the masonry and pointing absorbs water, thereby allowing the thin finish coat to dry out evenly and without cracking.

A professional plastererer will plaster a wall in one continuous operation, gauging the thickness they are applying as they work. However, the amateur plasterer will find it easier initially to divide the wall surface up into a series of bays using slim timber battens called grounds. These act as depth guides to help you apply an even thickness of plaster to each bay. They are removed once the plaster has set, and the narrow channels are then filled with more plaster, ready for the finish coat to be applied.

tools for the job

tape measure
tenon saw
claw hammer
spirit level
bucket
power drill & mixer attachment
spot board
workbench
garden spray gun
hawk
plasterer's trowel
stepladder
wooden rule
(1.5m/5ft length of 75 x 25mm/ 3 x 2in planed softwood)
wooden plasterer's float
angle trowel

1 Nail 25 x 10mm (1 x ³⁄₈in) planed timber battens to the wall you are plastering at roughly 1m (3ft) intervals, using 25mm (1in) long masonry nails. Check that they are vertical using a spirit level, and insert cardboard or hardboard packing behind them if the wall surface is uneven. Fit a batten right in the angle of internal corners, and pin on a length of expanded metal angle bead at external corners – this acts as a depth guide during plastering, and remains in place when the wall has been plastered to reinforce the corner.

2 Mix your first batch of plaster. As a guide to quantities, 50kg (110lb) of undercoat plaster will cover around 8sq m (86sq ft) of wall surface in a layer about 10mm (³⁄₈in) thick. Half-fill your bucket with clean water, then sprinkle handfuls of dry plaster into it, stirring as you do so by hand or with a power drill and mixer attachment. Add more plaster until the mix takes on the consistency of porridge. Tip it out onto your spot board, which should be set on a portable workbench, close to the wall you are plastering.

3 Wet the surface of the masonry in the first bay using a garden spray gun. Wetting the masonry will cut down substantially the absorption rate of the wall, thus preventing moisture from being sucked out of the undercoat too quickly, which will result in poor adhesion and cause the plaster to crack.

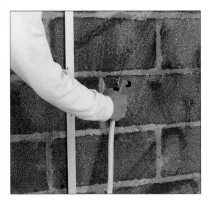

4 Hold your hawk under the edge of the spot board and scoop a trowelful of plaster onto it. Then transfer the plaster off the hawk and onto your trowel, tilting the hawk so that it is nearly vertical as you slide the almost horizontal trowel upwards across its surface. This procedure may take a little practice.

5 Resting the right-hand edge of the loaded trowel against the right-hand batten at floor level, tilt the blade until it is at an angle of about 30° to the wall. Push the trowel upwards to press the plaster against the masonry, gradually tilting the blade towards the vertical so that the plaster is squeezed out between its lower edge and the wall. The blade will be vertical when the plaster runs out. Load and apply a second band of plaster to the left of the first. Work your way up the bay to the next batten, applying parallel bands of plaster and blending them together. Over-fill the centre of the bay so that the plaster is proud of the battens. Use steps to reach the top of the bay.

6 When the bay is complete, hold a wooden rule across the guide battens at floor level and slide it upwards, at the same time moving it from side-to-side with a sawing motion. This will remove any high spots from the plaster. Fill obvious low spots with more plaster and then drag the rule across the bay again.

7 Drive five or six wire nails through a wooden float in a line 25mm (1in) in from one end of the blade, so that they protrude about 3mm (⅛in). Use this to key the surface of the fresh plaster. Wet the base of the float and hold it flat against the plaster. Then move it around in a circular motion so that the nails score shallow marks.

8 Repeat steps 3 to 7 to apply undercoat to the remaining bays. Prize out the battens and fill the channels with plaster. Use an angle trowel to plaster internal angles. Plaster external corners using a batten and the angle bead as depth guides.

9 Undercoat plaster takes about two hours to set hard, so you should be able to start applying the finish coat to the first bay by the time you have completed the undercoat for the final bay. Mix up a quantity of finishing plaster to the consistency of melting ice cream. Load your hawk and trowel as before, applying a coat with a maximum thickness of 3mm (⅛in). Start at the bottom of the wall and work upwards, again with broad sweeping arm movements. Cover about 2sq m (22sq ft), then apply another, even thinner, coat over the top of the first. Trowel off any ridges and splashes, and repeat the process until the entire wall is covered. Finally, wet the trowel and polish the finish plaster with the blade held flat against the surface.

MOVING BATTENS

As you become more proficient at plastering and gain in confidence, try using the moving batten technique. Put up two battens about 1.2m (4ft) apart and plaster the area between them. Then remove both battens and reposition one on the wall 1.2m (4ft) away from one edge of the plaster, which will act as a depth guide at one side of the next bay. Plaster this next section, then reposition the batten to the same distance again. Continue working around the room, bay by bay. When you return to your start point, the edge of the plaster in the first bay will act as your final depth guide.

working with plasterboard ⚒

Plasterboard is a rigid sheet material used for cladding the surfaces of timber-framed partition walls and for surfacing ceilings. It has a lightweight plaster core, sandwiched between two sheets of strong paper that also cover the longer edges of the board. Grey-faced plasterboard is intended for plastering over, while ivory-faced board can be painted or wallpapered directly. Ivory-faced boards with a tapered edge allow the joints to be taped and filled flush, ready for decorating.

types of plasterboard

Standard plasterboard, also known as wallboard, has one grey and one ivory face and is the most widely used type. It comes in standard 2.4 x 1.2m (8 x 4ft) sheets and a range of smaller sizes useful for repair jobs. You can also get longer sheets for rooms with high ceilings. Wallboard is available with square or tapered edges, in 9.5 and 12.5mm (³/₈ and ¹/₂in) thicknesses.

Baseboard is used for lining ceilings, and will be given a skim coating of finish plaster. It is available only as a square-edged board and has grey paper on both faces. The most common size is 1.2 x 0.9m (4 x 3ft). It comes in just the 9.5mm (³/₈in) thickness.

Both types are available with a vapour barrier. Vapour-shield boards are used mainly for upstairs ceilings (to prevent condensation in the loft) and for dry lining exterior walls (see pages 100–1). Thermal board has a layer of rigid insulation bonded to one face and also incorporates a vapour barrier. It is used for walls where extra insulation is needed.

storing plasterboard

Plasterboard is fragile until it is fixed to a supporting framework. Always carry sheets on edge – they may snap if you carry them flat. Store them on edge, closely packed against each other and leaning against a wall at a slight angle. If you are using ivory-faced boards, stack them with these faces together. Take care not to damage the paper-covered edges as you handle them.

fixing plasterboard

To line a partition wall or ceiling, plasterboard sheets are nailed to supporting timbers – vertical studs and horizontal top and bottom plates in a timber-framed wall, and joists in a ceiling. These are usually positioned at 400mm (16in) centres so that the edges of 1.2m (4ft) wide boards can be butt-jointed over the centre of every third timber and nailed to the intermediate ones. They are fixed with galvanized plasterboard nails, which have a jagged shank to grip the timber and a flat head that should be driven so that it dimples the face of the board. These are then filled with plaster to conceal them. Nails should be placed every 150mm (6in), 9mm (³/₈in) in from paper-covered edges and 12mm (¹/₂in) from cut ends.

cutting plasterboard

You can cut plasterboard with a fine-toothed saw, resting it on trestles to leave the cutting line clear. However, it is easier to cut through the paper and into the plaster core along the cutting line and then to snap the board over the edge of a length of wood. Cut through the paper on the other face to separate the two pieces. Use a padsaw, jigsaw or knife to make cut outs for light switches, socket outlets and so on.

MAKING A FOOT LIFTER

A foot lifter is a double wedge used to lift plasterboard sheets tight against the ceiling. Make one from a short length of 75 x 50mm (3 x 2in) wood, tapered into a wedge shape from the centre towards each end so that it rocks like a seesaw. Rest the sheet on one end of the wedge, then press down on the other end with your foot to lift the board into position. The small gap at the bottom of the board is concealed with skirting boards later.

cladding a partition wall

With the wall framework in place, fix the first board beside the doorway if the wall has one, as in this example, or in a corner if not.

tools for the job

tape measure
pencil
long straight edge
fine-toothed saw
trimming knife
foot lifter (see box above)
hammer

1 Measure the floor-to-ceiling height and subtract 20mm (³/₄in), then cut the board to length. Offer it up to the frame with one edge aligned with the door stud, lift it tight against the ceiling using a foot lifter and mark the position of the door head on this

edge. Cut a 25mm (1in) wide strip off this side of the board between the mark and the top edge. This cut edge will be centred on the upper section of the stud above the door opening.

2 Set the board back in place and nail it to the framework. Repeat the process for the board at the other side of the door opening.

3 Fix more whole boards in place, working from the doorway towards the corners. Butt tapered-edge boards together, but leave a 3mm (⅛in) gap between square-edged boards (to be plastered later).

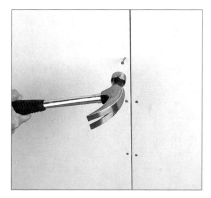

4 Cut the last boards down in width to fit the space at the room corners, and nail them in place. See pages 98–9 for how to fill joints and plaster plasterboard.

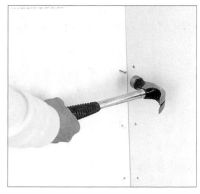

When cladding a ceiling, the boards should be fixed with their long edges at right angles to the joists, and with board ends meeting at the centre of a joist. To support the long edges, fix 50mm (2in) thick supports – called noggings – between the joists along the side walls, and across the room at centres to match the board width. You will also need steps or a platform of scaffold boards to work from, plus a spare pair of hands to help support the boards while each is fixed.

1 Offer up the first board in one corner of the room. Nail it to the joists and to the noggings, working from the centre of the board outwards. This stops the board from sagging as you fix it.

2 Complete the first row of boards, trimming the last one to fit if necessary. Butt-join tapered-edge boards, but leave a 3mm (⅛in) gap between square-edged ones.

3 Start the next row with a board trimmed to reach the centre of a joist. This avoids having all the joints between boards aligned along just one joist. Continue in this way.

4 Finish the ceiling with a row of boards cut down in width to fit between the last row of boards and the wall. See pages 98–9 for how to fill joints and plaster plasterboard.

plastering plasterboard ⸓⸓⸓

Plasterboard can be given a thin overall coat of board finish plaster if it is fixed with the grey side facing outwards. If tapered-edge boards are fixed ivory side out, however, only the joints need filling before the boards are painted or papered. In both cases it is vital to tape all the joints first, using proprietary paper joint tape or self-adhesive joint mesh, to avoid the risk of the joints opening up in the future due to movement of the ceiling structure.

plastering & lining

98

safety advice

The step-by-step sequences on these two pages demonstrate how to apply a plaster or taped finish to a plasterboard wall. Exactly the same techniques are used to finish a plasterboard ceiling, with the only difference being that you are working above your head. This has certain safety implications, however, since it is essential that you have a safe working platform so that you can reach the ceiling surface comfortably and without stretching. Scaffold boards or staging set on adjustable trestles, or a low mobile work trolley on lockable castors, are all ideal. Check out what your local plant hire company has available.

tools for the job

stepladders & boards
for working platform

bucket & mixer

spot board

plasterer's trowel

hawk

scissors for joint tape
or joint mesh

angle trowel (optional)

garden spray gun

filling knives

close-textured sponge

plastered finish

Set up a working platform, ensuring that it is steady (see pages 36–7), and mix some finish plaster in a bucket to the consistency of melting ice cream.

1 Apply a thin band of finish plaster with a trowel along each joint line, then cut joint tape to the required length and bed it into the plaster band using the end of the trowel to press it into place. If you are using self-adhesive joint mesh, dispense with the plaster band and stick the tape directly onto the board surface. Repeat for all the joints.

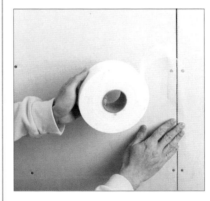

2 Spread a thin layer of plaster along each joint, wide enough to cover the tape or mesh. It should also be just thick enough to hide the tape or mesh completely. Smooth the plaster out on either side of the joints with the trowel held flat against the surface of the plasterboard.

3 Repeat steps 1 and 2 to bed tape or mesh into the wall/ceiling angle all around the room. Bed tape or mesh into the external and internal corners of the room in the same way. Do not be tempted to omit any of these angles – if you do not tape these joints, you are sure to get cracks opening up there as time goes by.

4 Apply finish plaster to the bays between the joints, working from the bottom upwards if you are plastering walls and from one edge outwards if you are plastering a ceiling. Use the same technique as for applying finish plaster over undercoat (see pages 94–5).

5 Return to your starting point and apply a second, thinner coat of plaster over the entire surface. Work with the trowel held almost flat to the surface to control the plaster thickness and ensure a flat finish.

6 Neaten the angles between wall and ceiling and in internal corners by running the edge of the trowel along each surface in turn. Alternatively, use an angle trowel.

7 Wet the blade of the trowel and the plaster surface using a garden spray gun, and polish the surface of the plaster until smooth.

taped finish

Use joint filler instead of finish plaster to fill the joints between tapered-edge boards. You can use either paper or self-adhesive mesh tape – the former is bedded in a band of joint filler, the latter is stuck straight to the plasterboard surfaces.

1 If you are using paper tape to fill the plasterboard joints, apply a narrow band of filler down the joint line and bed the tape into it with a filling knife, making sure that you exclude any air bubbles. Then apply another band of filler over the top with a wider knife to fill the tapered edges of the joint flush with the board surfaces at each side.

2 If you are using mesh tape to fill the plasterboard joints, stick the tape in place and fill the joint using the same technique as for paper tape. If you have to make a joint in the tape, butt the ends rather than overlapping them.

3 Finish all the joints with a thinner, wider band of filler, applied with a plasterer's trowel or a coating knife that will bridge the tapered edges of the boards. Then smooth out the edges of the filler with a slightly damp sponge.

PLASTERBOARD FINISHES

• **Sealing plasterboard** – To even out differences in porosity between the board and the filled joints, you should seal the wall before it is decorated, to prevent paint or wallpaper paste from drying too quickly as water is sucked into the board surface. To seal the plasterboard, use a thin coating of joint filler, applied and rubbed into the surface with a sponge. Alternatively, apply proprietary plasterboard primer with a brush or roller, or use emulsion paint diluted with 10 per cent water. Use two coats of paint to ensure uniform surface porosity.

• **Tiling plasterboard** – Plasterboard has very little strength if it gets wet, which can happen if a plasterboard wall is covered with ceramic tiles and the grouting is not waterproof. If you intend to tile an existing plasterboard wall, treat its surface with a coat of solvent-based paint first to seal the surface against water penetration. If you are cladding a new wall and plan to tile it, use special waterproof tile backing board instead of plasterboard to provide a surface that can withstand water penetration.

dry lining walls ⚒⚒⚒

This technique involves lining external masonry walls with plasterboard as an alternative to traditional plastering. It is used mainly in older properties to improve the insulation performance of solid walls. The boards are nailed to a framework of sawn timber battens fixed to the wall surface at centres to suit the board widths. Standard vapour-shield wallboard is commonly used for this purpose, with glass fibre insulation batts sandwiched between the boards and the masonry.

Dry lining can be installed over existing plaster if this is sound, but crumbly or damp plaster should be removed and, if damp, the problem must be treated and the wall allowed to dry out before dry lining is installed. The space between the dry lining and the masonry can be used to conceal cable runs to switches and sockets.

The timber used for wall battens should be pre-treated with wood preservative. It should be 50 x 50mm (2 x 2in) if glass fibre insulation is to be placed behind the plasterboard, and 50 x 32mm (2 x 1¼in) if thermal board is being used. Fix the battens with masonry nails long enough to penetrate the masonry by at least 25mm (1in), or with nail wall plugs.

If you are using insulation batts, wedge them in place between the wall battens all around the room before you start cutting and fixing plasterboard. Insulation is not used at the sides and heads of door and window openings.

Alternatively, thermal board with a layer of polystyrene or polyurethane insulation and a vapour barrier bonded to the rear face can be used to combine wall lining and insulation.

tools for the job

tape measure & pencil
straight edge
hand saw or power saw
spirit level
plumb line
cordless drill/driver
foot lifter (see box page 96)

1 Cut the vertical battens to a length about 150mm (6in) less than the room height and fix them in place, leaving a 75mm (3in) gap above and below each batten. Space the battens at 400mm (16in) centres for 9.5mm (⅜in) thick plasterboard in sheets 1.2m (4ft) wide, and at 600mm (2ft) centres for 12.5mm (½in) thick plasterboard in sheets of the same width. Start fixing the battens at door and window openings and work out towards the room corners.

2 When fixing battens at the corners of the room, fix a batten on each wall about 50mm (2in) away from the internal angle.

3 Cut and fit the horizontal battens at floor and ceiling level, fixing them in the gaps that you left above and below the vertical battens.

4 Add short horizontal battens above doorways and above and below window openings. Lastly, fit short vertical battens above these openings, offset by 25mm (1in) so that they will support the edges of boards that are fixed flush with the edges of the opening below as well as the infill panel above the opening.

5 Start fixing boards beside a door or window opening if there is one, and in a corner otherwise.

Cut the board to the required length, then use a foot lifter to hold it tightly up against the ceiling. Once it is in the correct position, drive in the fixings. Repeat the process at the other side and above the opening.

6 Cut strips of plasterboard to the required size to line the sides and head of the reveal opening around any windows or doorways. Place dabs of plaster on the reveals.

7 Fix the boards to the sides of the reveal so that their paper-covered edges lap the cut edges of the wall lining.

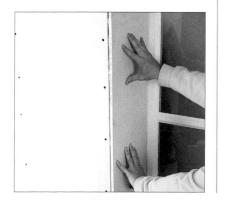

8 Prop the piece lining the head of the reveal in place with wooden battens until the plaster sets.

9 Work towards each corner of the room, fixing as many whole boards as possible. Cut the last board 20mm (³⁄₄in) narrower than the distance between the last whole board and the next wall. Fix it in place with its cut edge facing the corner.

10 Start the next wall with a whole sheet, butting the board against the face of the board already in place (but repeat from step 5 if there is an opening in the wall).

11 At external corners, fix one board with its cut edge flush with the face of the batten on the other wall. Then fix a board with a paper-covered edge to this wall so that the edge laps the cut edge of the first board. If using thermal board, you will have to cut away some of the insulation from the back of the board to allow board-to-board contact.

tips of the trade

If there are any switches or socket outlets on the wall that you are dry lining, make cutouts in the plasterboard to fit around them. If these are flush-mounted, remove the faceplates after turning the power off at the mains and draw the cable forward through the cutout. Fit a plastic cavity wall box in the cutout, feed in the cable and reconnect the faceplate to the new box.

FIXING BOARDS WITH ADHESIVE

If the existing wall surfaces are flat and true, you can fix ordinary plasterboard directly onto the wall surface using panel adhesive applied with a cartridge gun. Thermal board can be fixed with special adhesive applied to the wall in bands behind the centre and edges of the boards. This method reduces slightly the amount of floor area lost by dry lining a room, and saves the cost and time of fitting battens.

fitting an arch former ⁄⁄⁄

Prefabricated arch formers make it easy to convert doorways, alcoves and the openings between through rooms from plain rectangles to elegant arched openings. The formers are made from expanded metal mesh panels that are assembled to make arch shapes with flat vertical faces and curved undersides. The external angles are formed from the same metal bead that is used to reinforce external angles in plastering.

In the simplest kits, each end of the arch consists of two sections in the shape of a quarter circle. You offer them up against the masonry on each side of the wall. On thin walls, the two soffit sections overlap, while on thicker walls the gap between them is bridged by a soffit strip that is wired into place. Used at each side of a doorway, they form a semicircular arch. More elaborate kits contain shaped side and centre pieces to allow you to create a full-width flat or oval arch across the opening between two through rooms. These mesh kits are not suitable for fixing to timber-framed walls.

Choose an arch former to suit the opening you wish to convert, and check that it can cope with the thickness of the wall to which it will be fitted. Unpack the components and read the instructions carefully. You will need some special metal lathing plaster for the undercoat on the arch former, and standard finish plaster for the finish coat.

tools for the job

pencil
brick bolster
club hammer
claw hammer
wire cutters
hacksaw
pliers
screwdriver
bucket & mixer
hawk
plasterer's trowel

1 Hold the end sections of the former in position and mark the top and side edges on the flanking wall. This indicates the area of plaster to be removed. If you are installing a wide arch, repeat the process for the centre sections. Use a club hammer and brick bolster to chop away the plaster inside the marked lines beside and above the opening.

2 Hold the first side former section in place against the bare masonry, and secure it in place with masonry nails or screws and wall plugs, according to the manufacturer's recommendations. Check that it is level and square to the edge of the opening.

3 Fix the reverse section of the former to the other side of the wall, checking that it is accurately aligned with the first section that you fixed in step 2. Make sure the fixings are driven in flush with the mesh. Cut short pieces of galvanized wire and use them to tie the two soffit sections together, twisting the ends around each other neatly with pliers.

4 If the wall is thicker than about 215mm (8½in) – one brick length – the two soffit sections will probably not meet. In this case, cut a strip of the soffit mesh supplied with the kit and use it to link the two sections, again using wire twists to hold it in place.

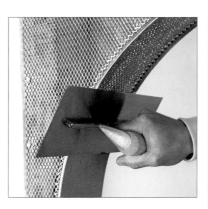

5 Repeat the process to fit the arch former sections at the other side of the opening, again adding an extra soffit strip if the wall is wide. Tuck the ends of the wire twists back through the mesh so that they cannot snag your trowel when you start plastering the arch.

6 Add any centre sections of a wide arch, fixing the former sections to the masonry above the opening and wiring the soffit together. Some kits include small plastic inserts to connect the metal beading together where the sections of former meet. Fit these before fixing the arch sections in place.

7 Load a hawk with some metal lathing plaster and start plastering the underside of the arch at the bottom of one side. Use the curved angle beads as a depth guide and press the plaster firmly into place so that it squeezes through the mesh and forms a strong key. Work up to the top of the arch, then plaster the other side from the bottom upwards.

8 Once the underside of the arch has been plastered, you should then tackle the side sections at each end of the arch, using the angle beads and the edge of the existing plaster as your depth guide. Score the plaster surface with the corner of the trowel to key it ready for applying the finish coat. Trowel on a thin coat of finish plaster as soon as the undercoat has dried hard. Tackle the soffit first, then the side sections, feathering out the plaster where it meets the existing plaster for an invisible joint.

ARCHES IN STUD WALLS

If you want to form an arch shape in an opening in a timber-framed wall, make up a simple timber framework at each side and fix plasterboard pieces cut to the shape you want. Then cut a strip of expanded metal mesh to fill in the soffit and pin it to the framework. Plaster this with metal lathing plaster, and tape the joints between the new and existing plasterboard before skimming the new board with finish plaster.

Take time applying the finish coat of plaster, making sure that you apply it smoothly and feather the edges where it meets the existing plaster to create an invisible join.

fitting coving ⌐

Coving is a decorative moulding that is fixed into the angle between wall and ceiling, framing the ceiling in the manner of a picture frame. It is generally a plain quadrant in cross-section, and may be made from plaster with a paper cover, or from a foamed plastic resin. More elaborate mouldings are generally referred to as cornices and have three-dimensional surface designs. They are traditionally formed in fibrous plaster but foamed plastic types are also available.

Coving and cornice is usually sold in lengths of 2 or 3m (2 or 3yds), which are butt-jointed along each wall of the room. It is mitred at internal and external corners, although some moulded plastic cornice has pre-formed corner pieces. Plasterboard and plastic coving are fixed in place with adhesive, but the weight of fibrous plaster mouldings makes fixing with screws and wall plugs essential. Old wallcoverings must be removed from the area beneath the coving so that the adhesive can bond to solid plaster or plasterboard. There is no need to fill cracks in the wall-ceiling angle – the coving will hide these.

fixing plasterboard coving

Measure the perimeter of the room and calculate how many lengths of coving will be required. Allow for the loss of about 150mm (6in) of coving for every mitre joint. Buy enough adhesive to fix the quantity of coving. Plasterboard coving is fixed with either powder or ready-mixed tub adhesive, while plastic coving usually has its own special adhesive.

tools for the job

tape measure
pencil
long straight edge
spirit level
trimming knife
wallpaper scraper
filling knives
mitre box (see box below)
fine-toothed saw
stepladders
hammer

1 Hold a length of coving in place in the wall/ceiling angle and mark pencil lines on the ceiling and wall along its top and bottom edges. Use a straight edge (plus a spirit level for the wall lines) to extend these lines around the room. If there is wallpaper on either surface, cut along the pencil lines with a trimming knife and dry-strip as much wallpaper between the lines and the wall/ceiling angle using a scraper. Soak any patches that remain to soften the paste, but take care not to wet the wallpaper outside the guidelines.

2 Use the edge of a filling knife to score the plaster surface with a series of criss-cross cuts in between the guidelines on the wall and ceiling. This will help the coving adhesive to bond well to the plaster and thereby achieve a secure enough fixing to hold the coving in place.

3 Start work on the longest wall in the room. Refer to the guidelines for cutting mitres on the opposite page to ensure that you cut the first mitre in the right direction – mistakes are wasteful and expensive. Place the coving in the mitre box with the ceiling edge in the base of the box, and cut with a fine-toothed saw. Smooth the edges with sandpaper.

MAKING A MITRE BOX

If you cannot find a mitre box that will cope with standard 100 and 125mm (4 and 5in) plasterboard coving, make one from scrap wood using a combination square to mark the 45° cutting lines. Cut the saw guides right down to the base of the box so that you can saw the mitres cleanly.

4 Use a wide filling knife to butter a generous amount of adhesive along the rear faces of the coving. Draw the knife towards the outer edge of the coving so that excess adhesive will mainly be squeezed into the triangular space between the coving and the wall-ceiling angle.

5 Align the bottom edge of the coving with the pencil line on the wall, slide the tip of the mitre into the corner and push the coving firmly upwards and backwards to compress the adhesive and ensure a good bond to the wall. This is a job for two people if you are working with 3m (3yd) lengths of coving. Scrape off excess adhesive along the wall and ceiling edges of the coving and check that it is lined up with the pencil lines. The adhesive should grab strongly enough to support the weight of the coving, but if you want to reinforce it while it sets, drive two or three masonry nails through its lower section and into the wall. Only drive them part-way in so that you can remove them and fill the holes later.

6 Measure the distance from the square end of the first length to the next corner. If it is more than one length away but less than two, cut a mitre on the next length to fit the corner and stick it in place as in step 5. If it is less than one length away, cut the mitre as before, then hold the length in position and mark where it meets the first length. Cut the coving squarely at this point and fit it in place, filling the joint between the two lengths with coving adhesive.

7 If you need a short infill piece to bridge the gap left between two full lengths, measure the gap and cut a piece of coving to fit it. Fix in place, filling the joints with coving adhesive.

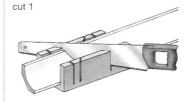

8 Move to the next wall and cut the correct mitre to fit into the corner, then repeat steps 3 to 7 to fix coving around the rest of the room. Finally, fill all joints with adhesive and remove any masonry nails you used to support long lengths of the coving.

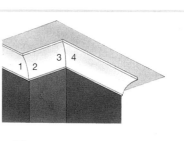

CUTTING MITRES

Take care that you cut mitres in the correct direction. The cutting diagrams below show the order of mitre cuts needed to cove a chimney breast (right), when working from left to right.

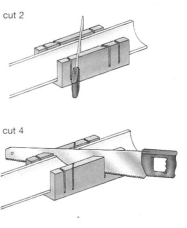

1 2   3 4

cut 1

cut 2

cut 3

cut 4

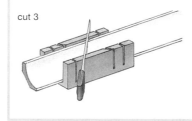

adding plaster features ↗

There are several other types of plaster feature that you can add to a room to complement the effect of coving (pages 104–5). These include decorative ceiling centres, wall plaques, corbels to support door heads, panel mouldings to frame areas of wall or ceiling, and even decorative plaster niches to display a favourite ornament. There is a wide range of styles available, from traditional to modern, formed in fibrous plaster or foamed plastic resin.

Ceiling centres and roses are widely available. They are produced in a wide range of different styles to suit every budget and taste, from traditional ornate examples to simpler, more modern varieties. Their function is twofold – they provide visual adornment to a ceiling and also conceal the electrical connections to a pendant light or chandelier. The smallest, lightest types can be stuck to the ceiling in the same way as coving, but large centres – especially those made from fibrous plaster – need to be screwed to the ceiling joists.

Wall plaques, being smaller and lighter than ceiling centres, are generally stuck in place, as are panel mouldings. Corbels fitted to support a door head or flat arch between rooms need screws driven into wall plugs, while plaster niches are generally fixed using mirror plates or similar fixings. Check the manufacturer's instructions when you choose the plaster feature you plan to install.

tools for the job

tape measure

string

drawing pins

pencil

wide filling knife

wallpaper scraper

bradawl

power drill & twist drill bits

screwdriver

fine-toothed saw

stepladder

fixing a ceiling centre

1 If your room has no central pendant light, the first step is to find the centre of the ceiling – a ceiling centre will look odd if it is not carefully positioned. To do this, find the mid-points of opposite walls and pin string lines between them. Where they cross is the ceiling centre. With the string lines still in place, hold the ceiling centre over them so that you can position it precisely central. Draw a light pencil line on the ceiling around its perimeter, then take it and the string lines down. (See the sequence on the opposite page for what to do if there is a central pendant light.)

2 If the ceiling is painted, use the edge of a filling knife to score the surface inside the line in a criss-cross pattern and provide a key for the adhesive. If there is wallpaper present, cut along the pencil line and scrape off as much of the wallpaper as you can. Soak the remains with a wet sponge and scrape them off too. Then key the plaster as before.

3 If the ceiling centre you are fitting is lightweight to medium-weight, simply fix it in place with adhesive. Apply a generous band of adhesive all around the edge of the ceiling centre with a filling knife, and add a blob in the middle too. Offer the ceiling centre up to the marked outline and press it firmly upwards against the ceiling surface. Hold it for a few seconds to give the adhesive a chance to grab, then scrape off excess adhesive around the perimeter and use it as a filler in any gaps between the centre and the ceiling.

4 If you are installing a heavy plaster ceiling centre, follow steps 1 and 2 to position it, then locate the joist positions at either side of the ceiling's central point. Use a bradawl within the pencil outline to probe for solid timber, and mark the joist positions at the perimeter of the outline. Hold the ceiling centre in place and mark where to drill holes for screws that will pass into the centre of the joists – use at least two, but preferably four, screws.

5 Take down the ceiling centre and carefully drill countersunk holes through it at the marked points. Spread a generous layer of adhesive on the back of the ceiling centre as described in step 3, then get a helper to hold the ceiling centre back in place while you drill pilot holes into the joists and drive in the fixing screws. Press the ceiling centre tightly against the ceiling as you do this, but take care not to overtighten the screws or you may crack the plaster.

6 Fill any gaps around the edge of the ceiling centre with adhesive, then wipe away the excess with a damp sponge for a smooth finish.

7 Finally, fill the fixing screw holes with a little adhesive and sand it smooth when it has set hard.

coping with a pendant light

1 Turn off the power at the mains and disconnect the pendant flex from its terminals, making a sketch of which cable cores are connected to which terminal on the baseplate first. Poke the wires up through the hole in the ceiling. Gain access to the ceiling void and connect the cables to the terminals of a four-terminal junction box, referring to your sketch to see which cables go to which terminal. This will restore the mains supply and switch control to the light when the power is restored later. Screw the

safety advice

Always consult a qualified electrician if in any doubt about carrying out wiring and electrical work yourself.

base of the junction box to the side of a nearby joist. Drill a hole in the middle of the ceiling centre for the pendant flex, and fit the ceiling centre as described on the opposite page.

2 Pass the end of a new piece of cable down through the ceiling. Connect the cores to the live, neutral and earth terminals of the junction box. Fit the cover on the box.

3 Return to the room below and connect the cable to the new baseplate. Screw this to the ceiling centre, connect the flex and fit the cover. Restore the power supply.

repair & restoration

Masonry surfaces are by their nature relatively hardwearing, but walls, ceilings and solid floors can all be damaged by accidental impact, structural movement and disasters such as plumbing leaks. If they are outdoors, the weather adds another factor, damaging brickwork and rendering, cracking concrete and causing paving to subside. Once the exterior of the house stops being waterproof, damp starts to appear on interior surfaces. Tackling all of these problems is well within the capabilities of most home-owners, as most are labour-intensive low-cost jobs that contractors will charge high prices to put right. Deal with them as soon as they develop, and you will stop a small problem turning into a bigger one.

If you are renovating a room it is vital to restore the plasterwork before decorating, especially if the walls are to be painted.

making minor repairs to plaster ↗

Plastered wall surfaces can be damaged by accidental impacts that leave dents and scrapes, and may develop other defects with age. Ceilings are less prone to damage, but can suffer from popped nail heads and from cracks developing along board edges, at the wall/ceiling junction or in line with the laths in an old lath-and-plaster ceiling. All can be repaired quickly and easily, and are well within the capabilities of even those who are less experienced in DIY.

tools for the job

nail punch

hammer

pliers

filling knife

trimming knife

old paintbrush

caulking blade

coating blade

belt sander

ceilings

The two main causes of ceiling defects are vibration caused by traffic on the floor above, and movement in the timber structure that supports the ceiling surface. Trouble can also arise if the joints in a plasterboard ceiling were not properly sealed when the ceiling was first put up.

popped nail heads

If you have plasterboard ceilings, you may find occasional tell-tale discs of plaster on the floor – evidence that one of the nails used to fix the boards to the joists has popped and dislodged the plaster skim covering it. This happens if the nail was not driven fully in when the ceiling was put up, so that the boards have moved slightly and loosened it.

1 Use a nail punch and hammer to drive the nail in until its head just dimples the surface of the plasterboard. If it has popped far enough for you to grip the head with pliers, pull it out and drive it in again just next to the original hole. Make sure its head is indented in the plaster skim.

2 Apply a little ready-mixed filler over the nail head with a filling knife, leaving it slightly proud of the ceiling surface. When it has set hard, sand it down flush and apply paint over the patch.

cracked ceilings

Cracks in plasterboard ceilings generally follow the edges of the plasterboard sheets, and are caused by movement in the ceiling structure as temperature and humidity changes. In lath-and-plaster ceilings, the cracks may be irregular or run parallel to the laths to which the plaster is bonded. See pages 118–19 for larger-scale repairs.

1 Draw the blade of a trimming knife along the crack, undercutting each edge slightly so that the filler will be locked in place when it sets. Then use an old paintbrush to brush any dust from the crack. On lath-and-plaster ceilings, brush some water along the crack to stop the dry plaster from sucking moisture out of the filler and making it crack as it sets.

2 Load up a filling knife with filler and press it well into the crack, drawing the blade across it as you work. After filling a short section of crack in this way, draw the knife blade along the crack to smooth the filler level with the surrounding surface. Carry on in this way until you have filled the whole crack. Allow the filler to set hard, then sand the repair smooth and redecorate to conceal it.

👍

tips of the trade

If cracks open up repeatedly along plasterboard joint lines, apply self-adhesive mesh joint tape along them and cover the tape with a wide band of filler applied with a caulking blade (see joint cracks below). Alternatively, apply a flexible textured coating to bridge and disguise the cracks, or put up lining paper.

walls

The main problem with walls is damage caused by collisions with the plaster surface. The cause could be anything – carelessly moved furniture, boisterous children playing or general wear and tear. Solid walls can also suffer from hairline cracking in the plaster, while hollow walls can develop cracks along the joints between the plasterboard sheets. All are relatively easy to rectify.

dents & cracks

All you need to repair small-scale surface damage to solid and hollow walls is some filler and a filling knife, or a caulking blade for hairline cracks.

1 Remove any loose material from the damaged area using an old paintbrush. Then fill it slightly proud of the surrounding surface with filler, and allow it to set hard. Sand the repair flush with the surface using fine-grade glasspaper, then redecorate.

2 Where an area of wall is suffering from extensive hairline cracking, tap the surface with your knuckles to check that the plaster is still sound. If it sounds hollow, its bond to the masonry beneath has probably failed (see page 112–13 for how to repair it). If it is sound, use a caulking blade to apply the filler over the affected area, working it in different directions to ensure that all the cracks are filled. Allow it to set, then sand smooth as before.

joint cracks

Long, straight cracks often appear in plasterboard walls along the lines of the joints between adjacent plasterboard sheets – a sure sign that they were not taped before the plaster skim coat was applied. The best solution to this problem is to tape these cracks and skim a coating of filler over them.

1 Use a belt sander to sand off 1–2mm (¹⁄₁₆in) of plaster along the joint line, as wide as the sanding

belt. This ensures that the tape and filler lie flush with the wall surface when the repair is complete.

2 Stick a length of self-adhesive mesh joint tape down the cracked joint. Press the tape firmly into position, butt-joining lengths if necessary to cover the crack.

3 Use a coating or caulking blade to apply filler over the tape, running the blade down the wall at an angle so that the filler is finished flush with the surrounding wall surface. Allow it to set hard, then sand and redecorate for an invisible repair.

patching plaster ↗

If you discover areas of plaster on solid walls that sound hollow when tapped, or areas of plaster that have fallen away from the wall completely, the solution is to patch the affected area with new plaster. This is a relatively easy job to tackle, even if you have never used plaster before, because you have a solid base onto which to apply it, and the surrounding sound plaster to act as a guide to enable you to achieve a smooth, level finish.

tools for the job

club hammer

cold chisel

work gloves

safety goggles

old paintbrush

bucket & mixer

hawk

gauging trowel

plastering trowel

length of batten
to use as straight edge

1 Use a club hammer and cold chisel to chop away all of the unsound plaster back to a sound edge. It is important to wear gloves to protect your hands and safety goggles to protect your eyes from any flying chips of plaster. Make sure that you remove any plaster that remains stuck to the masonry within the area you are stripping.

2 Use an old paintbrush to remove the remaining dust from the hole, especially along the bottom edge where the majority of falling dust will collect. If this is not done, the new plaster will not bond properly to the masonry surface at the base of the hole and will eventually fail again.

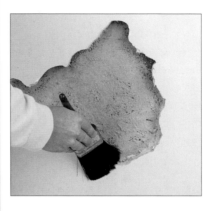

3 Dilute some pva (polyvinyl acetate) building adhesive with water, 1 part pva to 5 parts water, and brush a coat of this solution liberally onto the masonry, making sure that you reach right to the edges of the hole. This seals the surface, preventing it from sucking moisture out of the new plaster too quickly, which would cause the plaster to crack as it set. Allow it to dry before proceeding with the plastering.

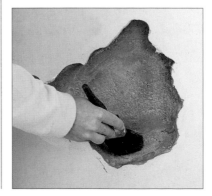

4 Mix some bonding plaster in a bucket until it is the consistency of thick porridge. Follow the manufacturer's guidelines regarding quantities, but remember that it is worth mixing a little more than you think you will need to ensure that you have a sufficient amount. Make sure that you stir the plaster thoroughly so that it is free from lumps – adding the plaster to the water rather than the other way around helps to avoid lumps in the mix. Scoop some of the plaster onto a hawk and use a gauging trowel to press the plaster into the hole. Start work at the edges of the hole, gradually filling the entire area to within 2–3mm (⅛in) of the surrounding surface.

5 When you have filled the hole evenly, lightly score the plaster surface with the tip of the gauging trowel in a criss-cross pattern. This process is called keying and is used to give the finish coat of plaster a better chance of bonding well to the underlying surface. Leave the undercoat plaster to set for a couple of hours before applying the finish coat.

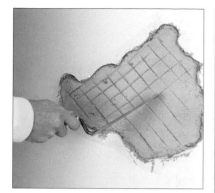

6 Mix a small quantity of finish plaster, this time to the consistency of melting ice cream. Once again, make sure you stir the plaster thoroughly to remove any lumps. Load some onto the hawk, scoop it off with the plastering trowel and apply it over the patch with a smooth, upward movement of the trowel. Tilt the blade towards the vertical as you finish the application to squeeze the plaster out between trowel and wall. Apply a second or third coat if necessary, with the end of the float resting on the surrounding plaster to help you finish the patch level with it.

8 Leave the repair to dry for a few minutes. Wash your plastering trowel clean, then use an old paintbrush to wet it with clean water, ready to give the repair its final polish.

tips of the trade

• **Ready-mixed plaster** – Unless you have a fairly large number of patches to repair, it is not usually worth buying separate supplies of bonding and finish plaster, since the smallest size generally available is 10kg (22lb). This quantity of bonding plaster is enough to cover about 1.5sq m (16sq ft); the same quantity of finish plaster will cover approximately 5sq m (54sq ft). It is therefore more economical to buy a tub of ready-mixed, lightweight, one-coat plaster and use this to fill the patch in one go. This type of plaster contains latex and other additives that are designed to prevent the plaster from slumping out of the hole when applied relatively thickly. Over-fill the patch slightly, leave the repair to set hard, then sand back with fine sandpaper until the repair is flush with the surrounding wall surface.

• **Storing plaster** – If you are using dry bagged plaster rather than a ready-mixed variety, store any leftover plaster in a tightly sealed plastic bag to prevent it from becoming damp and setting hard in the bag. Date the bag, and throw it away after six months if you have not used it by then – plaster does not keep well once opened.

9 Hold the trowel flat to the wall surface and use it to flatten and polish the patch to a smooth, hard finish. Flick more water onto the wall with the brush as necessary to keep the trowel wet, and do not be afraid to press hard with the trowel as you work. Leave the plaster to set hard, then sand off any flecks of plaster from the patch or the surrounding wall surface, ready for redecorating.

CHECKING FOR DAMP

One possible cause of failed plaster is dampness in the masonry. This may be the result of penetrating damp due to defects in the house structure – typical problems are water penetration around door or window frames, or roof defects allowing water to penetrate ceilings or down chimney breasts. Damp patches low down on downstairs walls may be caused by rising damp, due to defects in the damp-proof course. Whatever the cause, it is essential to put the defects right before patching the plaster, as continuing dampness will only cause the repair to fail again (see pages 134–5). If you are unsure as to whether a wall is damp, hire a damp meter from a local plant hire shop and use this to check all suspect areas.

7 Cut a piece of batten long enough to span the hole and use it as a ruler to scrape off any excess plaster that is sitting proud of the surrounding wall surface. This will also reveal if there are any hollow areas in your patch. If there are, apply an additional skim of finish plaster to fill them, then scrape the batten ruler across the patch once more to remove any excess.

repairing corners ↗

The most vulnerable parts of any plasterwork are external corners, especially in doorways where the angles are most likely to be knocked and damaged. Older homes are especially vulnerable, because the original plaster is often both thicker and softer than modern plasterwork and is also not likely to have been reinforced. Internal corners are less likely to suffer damage, which is generally limited to cracking in the angle due to differential movement in the house structure.

patching small-scale damage

If the damage is superficial, restoring the corner is a simple two-stage job. If the wall is papered rather than painted, first strip the wallpaper off in the region of the repair.

tools for the job

old screwdriver or small cold chisel
old paintbrush
tenon saw
power drill & masonry drill bits
hammer
filling knife

1 Chip away any loose plaster from the damaged area with an old screwdriver or a small cold chisel until the area has a clean, sound edge. Then brush away any dust from the area with an old paintbrush.

2 Cut a piece of batten twice as long as the damaged section and at least 50mm (2in) wide so that

the nails fixing it to the wall do not break away more plaster from the corner. Drill pilot holes near each end of the batten and push in a masonry nail. Hold the batten against one face of the corner with its edge flush with the other face and tap in the nails far enough to hold the batten in place. Leave the heads projecting so that you can remove them easily later.

3 Mix some filler and use a filling knife to pack it into the hole between the damaged edge and the batten on one face of the wall. Run the blade along the wall and batten surface so that the filler is flush with them. Press the knife in firmly to ensure that the filler bonds well.

4 Allow the filler to become touch-dry, then carefully remove the batten and reposition it on the other face of the corner, covering the section you have just filled and with its edge flush with the surface of the adjacent wall. Fill the rest of the hole as before, again running the knife on the wall and batten surfaces to leave a smooth, level finish to the repair.

5 When the filler has set, carefully prize off the batten and remove the nails from it. Fill the nail holes with more filler, then lightly sand the repair with fine-grade glasspaper, aiming to round off the repaired section slightly to match the profile of the rest of the corner angle.

repairing the entire corner

If the corner is extensively damaged, it is best to strip and repair the corner from floor to ceiling, incorporating a strip of metal angle bead beneath the repair plaster to make future damage less likely. Unless you have powder plaster available, buy a tub of ready-mixed repair plaster for this job.

tools for the job

tape measure
portable workbench
hacksaw
bolster chisel
club hammer
hawk
gauging trowel
plastering trowel
corner trowel (optional)

1 Measure the height of the corner. Clamp the angle bead in a portable workbench and cut it to length using a hacksaw. Saw through the bead first, then cut the expanded metal mesh wings that flank either side of the bead one at a time.

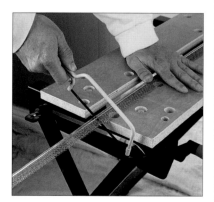

2 Use a bolster chisel and club hammer to chip off a band of plaster about 40mm (1½in) wide on each side of the corner. Draw pencil guidelines on each wall and cut along each line first, then remove the plaster from the corner.

3 Apply dabs of plaster at 300mm (12in) intervals on either side of the corner using a gauging trowel.

4 Press the angle bead into place until the plaster oozes through the mesh wings. The wings should almost touch the masonry to ensure that the bead is at the correct level.

5 Use the tip of the gauging trowel to flatten the extruded plaster over the mesh and to remove excess. Check that the bead is vertical and make any necessary adjustments. Allow the plaster to set for an hour before proceeding any further.

6 Next, fill the gap between the bead and the existing plaster. Work from the bottom up on one side of the corner, holding the plastering trowel at a 45° angle as you push it upwards and force plaster into the space between the bead and the existing plaster. Repeat the process on the other side of the corner.

7 Wet the float of the trowel with water and polish the repaired plaster to a smooth, flat finish. When the plaster has set hard, complete the job by using a damp cloth to wipe any plaster off the exposed quarter of the angle bead.

repairing hollow walls ↗

Most houses have some hollow walls with timber frames. In older houses, these were covered with slim timber laths that were then plastered over – a finish known as lath-and-plaster. In houses built since the 1930s, sheets of plasterboard have been used instead of lath-and-plaster. Both are prone to damage from impacts, which causes the plaster to break away from lath-and-plaster walls and generally makes a hole in plasterboard.

repairing lath-and-plaster

As long as the laths are not damaged, replastering is a straightforward job because the laths support the plaster. If any are broken, you need to provide support to prevent the plaster from falling into the void behind the laths.

tools for the job

trimming knife
old paintbrush
pencil
scissors or tin snips
cordless drill/driver
bucket & mixer
hawk
gauging trowel
plastering trowel

1 Use a trimming knife to cut away loose plaster from the edge of the damaged area, until it is surrounded by plaster that is still bonded to the laths behind. Poke any lumps of plaster trapped between the laths into the void behind.

2 Use an old paintbrush to remove dust and remaining debris from the hole and the laths. Take care that you dust the area thoroughly because loose material left in the hole will prevent the new plaster from bonding to the laths properly, and may lead to premature failure of the repair.

3 Hold a piece of fine metal mesh over the hole and draw a pencil line on it just inside the perimeter of the hole. Cut it to size with scissors or tinsnips, hold it in place and drill several slim pilot holes through the mesh and into the laths. Secure the mesh to the laths with short, slim countersunk screws.

4 Wet the edges of the hole with water to prevent the repair plaster from drying out and cracking. Mix some bonding plaster, and press it into the hole with a gauging trowel, working from the edges inwards. Fill the hole to within 2–3mm (⅛in) of the surrounding plaster. Score the surface of the wet plaster with the tip of the trowel in a criss-cross pattern to provide a good key for the finish coat.

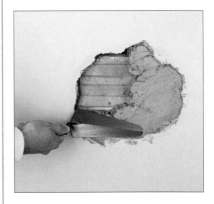

5 Allow the bonding coat to set hard, then mix some finish plaster and use a plastering trowel to skim-coat the patch and leave it flush with the surrounding wall. Allow it to harden, then wet the trowel and polish the surface to a smooth finish.

repairing plasterboard

If you have a hole in a plasterboard wall, there will be nothing behind it to support the repair plaster. The solution is to insert a piece of plasterboard or other board in the hole and bond it into place, then to fill the hole in the usual way.

tools for the job

pencil & ruler

trimming knife or padsaw

hand saw

cordless drill/driver

filling knife

1 Draw a square or rectangle around the hole and cut along the lines with a trimming knife. Use a padsaw to cut 12.5mm (½in) thick plasterboard, which is too thick to cut easily with a knife. Remove the cut section of plasterboard.

2 Cut a patch of plasterboard or other sheet material such as hardboard to the appropriate size – about twice the height of the hole in one dimension, and a fraction less than its width in the other. Drill a hole in the centre of the board and thread a loop of string through the hole. Tie the cut ends of the string around a nail and pull the string until the nail lies flat up against the face of the patch of board.

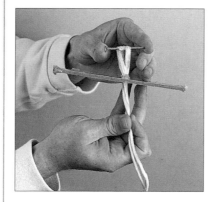

3 Apply some instant-grip adhesive to the other side of the patch along the two shorter edges. This will bond it to the inner face of the wall board when it is inserted in the hole.

4 Carefully feed the patch into the hole while holding the string in your other hand so that you do not lose it in the void. Manoeuvre it so that the two short edges press against the inner face of the wall board above and below the hole. Pull hard on the string to bond the patch in place. With instant-grip adhesive it

should stay safely in place after a minute or so. Allow it to set for the time recommended on the tube.

5 When the adhesive has set, cut off the string. Mix some filler and fill the hole in two stages – first to about half its depth and then, when this layer is touch-dry, to just proud of the surrounding surface. Allow it to set hard, then sand it flush.

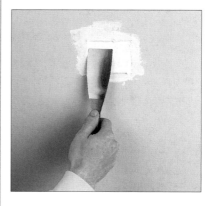

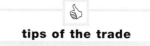

patching ceilings ⁊⁊

Ceilings can be damaged by water leaks, or sometimes on an old lath and plaster ceiling the plaster can simply come away from the laths if the plaster key fails with age, but the most common cause of damage to a ceiling is when someone's foot slips off a rafter onto an unboarded area of the loft and ends up poking through – the stuff of many a TV comedy sketch! After the comedy of the moment, however, there is no need for panic as a simple patch repair will return the ceiling to a finish as good as new. Whatever has caused the damage, the best solution is to cut away the affected area and to fit a plasterboard patch.

tools for the job

pipe, cable & joist detector
pencil
straight edge
safety goggles
padsaw
trimming knife
claw hammer
tape measure
panel saw
power drill & twist drill bits
screwdriver
self-adhesive mesh
joint tape
filling knife
plastering trowel
hawk

1 Locate the joists on either side of the hole, either by using an electronic joist detector or, if the loft is unboarded, by making holes down through the ceiling alongside the joists so that their positions are visible in the room below. Draw a pencil line along the centre line of each joist, extending beyond the damaged area.

safety advice

You must also check whether there are any services, such as water pipes or electricity cables, in the immediate vicinity of the hole before drawing guidelines for cutting. If the ceiling is below the loft these checks can be made manually, otherwise a cable, pipe and joist detector can be used.

2 Draw two further lines at right angles to the first, joining up the joist lines to form a rectangle outline of the area you plan to cut out. Use either a long ruler, a timber straight edge or a spirit level.

3 Cut out from the hole towards one of the pencil guidelines using a padsaw, then cut along the line in each direction until you reach the joists. Repeat the process at the other side of the hole. Use a trimming knife to cut the plasterboard along the guidelines indicating the joist centres. Look out for concealed fixing nails as you do this.

4 Pull down the cut-out section of board and use the claw of a hammer to prise out the old fixing nails from the undersides of the joists. Lever them out with the hammer head against the joist, not the plasterboard.

5 Cut two lengths of sawn softwood so that they form a tight fit between the joists on either side of the hole. These will act as noggings to support the edges of the plasterboard patch. They should be at least 50 x 50mm (2 x 2in) in cross section, but 100 x 50mm (4 x 2in) is ideal. Hammer them in until they are half concealed by the edge of the hole.

6 To secure each nogging in place you will first need to drill a clearance hole through one end at 45° to the joist, then drive in a long screw to secure the end of the nogging to the joist it butts up against. Repeat the process to fix the other end of the nogging to the opposite joist. Don't worry if the noggings move slightly as you screw them in place – there will still be enough exposed wood to which the plasterboard patch may be nailed.

7 Measure and cut a piece of plasterboard to fit the hole. Test its fit and, if necessary, trim the sides down to size with the aid of a trimming knife. Lift the piece of plasterboard into place and nail it to both the joists and noggings with 30mm (1¼in) galvanized plasterboard nails. Position the nails a distance of at least 10mm (⅜in) from the edges of the plasterboard, and drive them into the wood until their heads just dimple the paper. Nail the existing plasterboard to the noggings along the edges of the hole too.

8 Stick lengths of self-adhesive mesh joint tape over the joints to stop cracks opening up in the future. Then use a filling knife to apply a band of filler over the tape both to conceal it and to fill the gaps around the patch.

9 Plaster over the patch with a skim coat of finish plaster if you have it, or use a one-coat ready-mixed plaster. Feather out the edges of the plaster onto the surrounding ceiling surface to make the patch less noticeable. Wet the trowel and use it flat to the ceiling to polish the repair, leaving a smooth finish. Leave to dry and then decorate to match.

WATER DAMAGE

• **Plasterboard ceilings** – If a part of the ceiling has been soaked by penetrating damp or a plumbing leak, the surface finish will still be badly stained by the water even if the actual plasterboard has remained intact. It is not enough simply to reapply water-based paint as the stains will keep bleeding through. The solution is to seal them in with a proprietary stain-block aerosol, or to cover them with a coat of any solvent-based primer or paint. Once sealed you can then paint over the area with water-based paint to match the rest of the ceiling.

• **Lath-and-plaster ceilings** –
If the ceiling is of a lath and plaster construction, the technique for making the repair is slightly different:
1. Pull down as much loose material as you can to begin with, then mark out the area to be removed as for plasterboard ceilings.
2. Make the saw cuts parallel to the laths first, inserting the padsaw between adjacent laths and cutting through the plaster. Dislodge as much plaster as possible between these cuts and the hole to expose the laths that need to be removed.
3. Simply pull broken laths downwards so that they snap where they are nailed to the joists – they are usually very dry and brittle. Saw through the centre of any undamaged laths and snap them off in the same way.
4. Clean up the broken ends of the laths beneath the joists with your trimming knife, then cut and fit the plasterboard patch as normal.

If a patch of plaster breaks away because the key to the laths has failed, the rest of the ceiling may be close to failure too. If this is the case, the best solution is to pull down the entire ceiling and replace it with a plasterboard ceiling.

restoring plaster mouldings ⁄⁄

Many older properties have highly detailed plaster cornices and other ceiling and panel mouldings as original features. These would have been moulded in fibrous plaster and then screwed into place – most are too heavy to be supported solely by plaster adhesive. Unfortunately, years of repainting will often gradually obscure the fine detail. Many may also have been damaged by the building of partition walls to subdivide large rooms. However, restoration is usually possible.

cleaning mouldings

If you have mouldings that are clogged with layers of old paint, be prepared for some slow and fiddly restoration work. The first thing you have to do is to find out what sort of paint you have to contend with. In an unrestored property it will probably be distemper, but in renovated ones you could find anything from eggshell to modern emulsion paint.

tools for the job

work platform
garden spray gun
improvised picks & scrapers
old toothbrushes
soft-bristled brush

1 The first thing to try on old mouldings is water applied as a mist with a garden spray gun. This will soften distemper, but will have no effect on other types of paint. Soak a test area and leave it to penetrate for 10–15 minutes.

2 If the water works, use an improvised pick, scraper or toothbrush, as appropriate, to remove the old distemper bit by bit. This will be time-consuming and fiddly work, so tackle just a short section at a time and make sure you are working at a comfortable height – ideally off a work platform rather than a stepladder.

3 After removing as much paint as possible, scrub the surface of the moulding gently with a soft-bristled brush to remove flecks and specks left behind by the picks and scrapers. Repaint it with a thinned coat of emulsion to act as a sealer, followed by a full-strength coat.

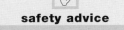

safety advice

Always wear PVC work gloves and safety goggles to apply any form of chemical paint stripper.

4 If water fails to make any impression on the old paint, you will need to experiment with chemical strippers. Look for products that come in paste form rather than as runny liquids. The former will stay put on the surface of the moulding as they soften the paint, while the latter will splash everywhere and make a thoroughly unpleasant mess. Brush the stripper on liberally, then work it into the recesses of the moulding with a stippling action of the brush.

5 Some strippers are designed to be used in conjunction with special fibrous tissue strips that you bed into the layer of stripper. These not only help to prevent the stripper from drying out too quickly, they also allow you to peel strip and stripper away in one go after the stripper has done its work, making the restoration job much less labour-intensive.

restoring damaged cornices

If the damage to the cornice is minor, you may be able to make it good with filler or plaster of Paris, moulding the repair material to match the originals. However, if this is not possible, the best option is to replace the damaged section with a new piece of cornice that matches the original as closely as possible.

There are several manufacturers who produce modern replicas of traditional cornices in fibrous plaster, so unless your particular cornice is very unusual you may be able to find a suitable replacement from one of these suppliers. Another possible source of matching cornices is architectural salvage companies, who rescue and sell period details from old houses. Both solutions are likely to prove expensive, however.

tools for the job

safety goggles

work gloves

brick bolster

club hammer

hacksaw

plastering equipment
(see pages 112–13)

cordless drill/driver

screwdriver

filling knife

paintbrush

1 If you are able to obtain a length of suitable replacement cornice to match the damaged cornice in your home, remove the damaged section by carefully cutting it away piece by piece with a brick bolster and club hammer. Remember to wear protective gloves and safety goggles to protect you from dust and flying debris. Depending on how the cornice was put up, you may find wall and ceiling plaster coming away as you work. Look out for fixing screws buried in the coving. Break the plaster away around them, then cut them off flush with the wall or ceiling surface with a hacksaw.

2 When you have removed the damaged section of cornice, clean up the newly exposed wall and ceiling surfaces so that the new cornice can fit closely against them. Cut the new cornice to length and offer it up to check its fit – this is a job for two pairs of hands. If the old cornice was put up after plastering, you will then have to replaster the areas that came away in step 1.

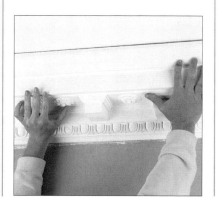

3 Make sure that any replastering has fully hardened before installing the length of replacement cornice. When it is ready, drill fixing holes through the cornice at the spacings recommended by the manufacturer. Hold the coving back in place and mark the corresponding hole positions on the wall. Drill the holes and insert wall plugs, then apply the adhesive recommended by the manufacturer to the rear faces of the cornice and offer it up to the wall and ceiling surfaces, pressing it firmly into place. Secure it with long screws countersunk beneath the surface of the cornice.

4 Insert filler into the screw holes and the joints between the old and new sections of cornice, then make good the wall and ceiling plaster alongside the repaired section. Finally, paint the cornice with a coat of diluted emulsion paint to seal the surface, followed by a full-strength coat. Brush the paint out well so that you avoid the clogging that overpainting can cause.

repairing a concrete floor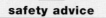

If you uncover a concrete floor that is showing its age, it pays to tackle any necessary repairs or renovation before putting down new floor coverings. If you do not, the problems will only get worse and can spoil the performance and appearance of the floor covering. You may have to seal it, patch it, damp-proof it or level it – or any combination of these. However, all such jobs are well within the abilities of any competent person, do-it-yourself enthusiast or not.

safety advice

Wear safety gloves and goggles and a face mask when working with concrete to prevent dust inhalation and injuries from flying debris.

sealing dusty concrete

One of the most common problems with concrete floors is dusting. This generally happens because the floor screed was over-trowelled when it was first laid, resulting in a surface rich in cement but low on strength. As time goes by, footfalls gradually cause the surface of the concrete to disintegrate, and when old floor coverings are lifted the surface is found to be covered in a layer of dust.

tools for the job

soft broom

dustpan

bucket

paint roller with extension pole

roller tray

1 Close the door and open the windows, then sweep the floor from one end to the other and collect the dust in your dustpan. Bag it and bin it. Do not sweep over-vigorously – you will simply create more dust as the surface of the concrete continues to break up. Leave the dust to settle for an hour before proceeding with the finishing step.

2 The next stage is to apply a sealing solution to the concrete. You can use a product called stabilizing solution, in which case simply pour some into the roller tray. If you choose pva (polyvinyl acetate) building adhesive, however, you should dilute it with water as recommended by the manufacturer, then pour this into the roller tray. Fit the extension pole and load the roller sleeve. Apply the sealer in bands, working back towards the door, and allow it to dry. Apply a second coat if the floor appears very porous. Using a roller is far quicker than brushing the sealer onto the floor, and an extension pole means you do not have to work on hands and knees.

patching concrete

If a floor has extensive cracks and small potholes, these will show through a smooth floor covering and are likely to get worse over time. Fix them now to avoid further detoriation.

tools for the job

cold chisel

club hammer

stiff brush

old paintbrush

bucket & mixer

hawk

pointing trowel

plastering trowel

1 Use a cold chisel and club hammer to break away any obviously loose material along cracks or around potholes. Continue until you reach a sound edge, then undercut this slightly to give the repair a chance to bond well with the surrounding masonry. Use a stiff brush to remove dust and fine debris.

2 Brush a coat of diluted pva building adhesive along cracks and inside holes. Then mix up some masonry filler or patching mortar and load it onto a hawk. Use a pointing trowel to force filler or mortar into the cracks, finishing off the repair flush with the surrounding floor surface.

3 Pack potholes layer by layer, then use a plastering trowel to give the patch a flat surface. When it has set hard, apply a coat of diluted pva building adhesive to seal it.

laying self-smoothing compound

An existing concrete floor will have been finished with a screed of fine concrete, usually about 50mm (2in) thick. If this is in generally poor condition, the best way of giving it a new smooth, flat surface is to cover it with self-smoothing compound. This sets to form a very thin, hard surface screed that is the perfect surface for all types of new floor covering.

tools for the job

stiff brush

paint roller with extension pole

roller tray

bucket

drill & mixing attachment

plastering trowel

1 Sweep the floor to remove loose dust and debris, then seal it with a coat of diluted pva building adhesive, applied with a paint roller (follow the method described in step 2 of 'sealing dusty concrete' on the opposite page). Allow the sealant to dry overnight.

2 Mix the self-smoothing compound in a bucket, following the manufacturer's instructions. These compounds are water-based, but may require the addition of one or more other chemicals to improve their workability and self-levelling ability. Whatever you need will be supplied with the compound.

3 Pour out enough self-smoothing compound to cover an area of about 2sq m (25sq ft), starting in the corner of the room that is farthest from the door. The compound will flow out freely and find its own level. Use a plastering trowel to spread it over the floor evenly up to the skirting boards. Then pour some more compound and trowel that out in the same way. Work back towards the door, area by area, then leave it to harden for about two hours before walking on it. You can lay new floor coverings over it the next day.

TREATING DAMP FLOORS

If you suspect that a concrete floor is damp – for example, because the old floor covering was stained or mildewed on the underside – hire a damp meter from a local tool hire shop. Use the meter to check the extent of the damp, which will probably have been caused by a failed damp-proof membrane in the floor structure of the house. Cure the problem by applying two coats of proprietary waterproofing emulsion, following the manufacturer's instructions about coverage and drying time between coats. Use it at the start of the day so that you can ventilate the room thoroughly while the compound dries. If you plan to use a self-smoothing compound after damp-proofing the floor, check that the two products are compatible when you buy them.

repointing brickwork 🔨

The mortar joints between bricks – the pointing – is the weakest link in any exterior wall. If it fails for any reason, water can penetrate the bricks and winter frosts can lead to their faces splitting off – a condition known as spalling. The problem may have been caused by poor workmanship, by use of the wrong mortar mix or simply by rain erosion. Make a point of inspecting your brickwork each spring, and tackle any areas showing signs of deterioration before serious problems arise.

tools for the job

safety goggles

work gloves

cold chisel

club hammer

stiff brush

bucket & mixer

garden spray gun

hawk

pointing trowel

profiling tools

1 If you find an area of pointing that is loose and crumbly, the first job is to chip it out back to sound mortar – to a minimum depth of about 20mm (¾in) – using a sharp cold chisel and a club hammer. Wear work gloves and safety goggles to guard against flying debris. Work along the horizontal joints first, then along the vertical ones.

2 Use a stiff-bristled brush – but not a wire one, which will mark the bricks – to clean dust and debris from all the joints you have worked on. Then mix a small batch of mortar

and allow it to dry to see how well it matches the existing mortar. Use a ratio of 1 part cement to 5 parts soft sand, and experiment with different coloured sands to get as close a match as possible.

3 When you have a suitable mortar formula, mix about half a bucketful at a time – repointing is slow work, and unused mortar that has begun to dry out cannot be resuscitated by adding more water. Then spray water from a garden spray gun onto the area where you plan to start work. This helps to cut the suction of the brickwork and prevents the mortar from drying out too quickly.

tips of the trade 👍

• **Tackling large areas** – If you have a large area of wall to repoint, you can speed up the chopping out process by hiring a power tool called a mortar raker. This has a tungsten carbide cutter that grinds out the old mortar to a pre-set depth in a fraction of the time the job takes by hand. Wear safety goggles, a dust mask and ear defenders when using this equipment.

• **Access equipment** – Repointing brickwork is a slow job, and working off steps or a ladder can make your feet and back ache. Work off trestles and staging instead for areas up to about 3m (10ft) above ground level, and use a slot-together platform tower for work higher up. Not only do these access options enable you to stand or kneel in comfort, they also provide a surface for placing tools and materials conveniently to hand. Both types of equipment can be hired.

4 Put some mortar onto a hawk, and take a sausage shape of mortar off it with a pointing trowel. Press the mortar firmly into one of the horizontal joints, and draw the tip of the trowel over it to bed it in and bond it to the underlying mortar. Repoint all the horizontal joints first.

5 Use the same technique to fill the vertical joints, one at a time. Press the mortar in well, leaving it almost flush with the surface of the bricks. Trim off excess mortar as you work, but leave any that gets on the face of the bricks to dry. You can then remove it with a dry brush and avoid staining the wall face.

6 When you have completed about 1sq m (10sq ft) of wall, it is time to finish the pointing to match the existing brickwork's joints. In a weathered joint, the pointing has a slope outwards from top to bottom, with the top recessed by about 5mm (¼in) and the bottom flush with the face of the brick below. Form this type of joint by drawing the tip of the trowel along the newly filled joint, with the flat of the trowel resting on the top edge of the brick below the joint.

7 To match a V-shaped recess, draw the tip of the pointing trowel along the centre of the new mortar joint, removing some mortar to leave a neatly shaped recess.

8 A joint with a concave profile is one of the most common joint finishes and is easy to match. Simply draw a rounded object such as an offcut of garden hose along the joint, leaving it with a semicircular cross-section. A recessed joint is set back from the face of the bricks by up to 10mm (⅜in). It is formed by drawing a wood offcut or similar implement along the joint to rake out the mortar to a uniform depth. This type of joint should not be used on brickwork in exposed locations.

tips of the trade

If you are having trouble matching the colour of your existing pointing, try adding a mortar pigment to your mix. These powders are available in black, brown, green, yellow and red, and come in 1.25kg (3lb) packs that will tint 50kg (110lb) of cement. Make a few trial batches first using different amounts of pigment, allowing them to dry out before you compare the final colour. Measure the ingredients accurately – too much or too little pigment will change the colour of the finished mortar noticeably.

Identify the type of joint in the existing brickwork and try to match it as closely as possible. The recessed joints in this brickwork are formed using an offcut of wood.

replacing damaged bricks ⁄⁄⁄

If failed pointing has allowed rainwater to penetrate behind the face of your brickwork, and subsequent frost has split off the faces of some of the bricks – known as spalling – the only way you can restore the appearance of the wall is to chop out the damaged bricks and replace them. The job itself is a relatively straightforward one. The biggest problem lies in finding replacement bricks that are a good match for your existing ones.

tools for the job

safety goggles

work gloves

cold chisel

club hammer

hammer drill & masonry drill bits

brick bolster

bucket & mixer

spot board

hawk

bricklaying trowel

pointing trowel

profiling tools

1 Chop out the pointing all around the damaged brick using a sharp cold chisel and a club hammer. Make sure that you wear work gloves to protect your hands, and safety goggles to keep flying debris out of your eyes.

2 To make it easier to remove the damaged brick from the wall, drill a series of closely spaced holes down the middle of the brick to a depth of about 100mm (4in) using a hammer drill and a large-diameter masonry drill bit. Take care that the drill bit does not slip onto any of the surrounding bricks and cause further damage.

3 Try to split the brick in half by chopping along the line of the drill holes with a brick bolster and club hammer. When you have done so, hold the bolster on this central split at an angle towards the ends of the brick and chop out sections one by one, taking care not to damage the surrounding bricks. Take care on cavity walls not to drive pieces of brick into the cavity, where they could drop and act as a damp bridge between the inner and outer parts of the wall.

tips of the trade

It can be difficult to remove whole bricks from solid walls, especially walls in which the bricks that are laid end-on (known in the trade as headers). The best way of repairing a damaged header is to drill it as described in step 2, and then to chop it out to a depth of about 100mm (4in) – roughly half the length of the brick. Chop the replacement brick in half and test its fit in the recess, cutting it down further in length if necessary to allow for a mortar bed behind it. Butter mortar into the base of the recess and onto the top, sides and back of the replacement brick, and slide the brick into place. Check that it is centred in the recess, adjust if necessary, and then tap the brick gently home until it is flush with its neighbours. Point all around it, matching the existing pointing as closely as possible, to complete the repair.

4 When you have removed the damaged brick, chop away as much of the remaining pointing as possible from the top, bottom and sides of the recess. Once again, if you are repairing a cavity wall, take care not to let any debris drop down into the cavity.

5 Mix a small amount of mortar and let it dry to see how well it matches the colour of the existing pointing. See the tips of the trade box on page 125 for more information on matching the colour of mortar. When you have a satisfactory formula, mix another small batch and place it on a spot board – a piece of old plywood or similar board close to where you are working. Place a bed of mortar on the base of the recess using a bricklaying trowel, and butter more mortar onto the top and ends of the new brick.

See the tips of the trade box on page 125 for more information on

6 Carefully slide the new brick into place in the recess, centring it and then tapping it backwards with the handle of the club hammer until it fits flush with its neighbours. Check that it is horizontal and that the joints around it are an even thickness, repositioning it slightly if necessary to achieve this. Use a pointing trowel to add mortar as necessary to the joints around the new brick, then finish the pointing to match the rest of the wall.

MATCHING BRICKS

Unless your house is relatively new, the brickwork will have weathered and changed colour over the years to the point where a new brick – even if it is a close match to the originals – will look highly noticeable next to its aged neighbours. Look out for advertisements in local papers offering second-hand bricks, which may be a closer match to yours. Architectural salvage firms may also be able to help. If all else fails, tour your area looking for demolition work in progress – you may well find the bricks you want lying in a skip somewhere.

Once you have found an appropriate second-hand brick, you will need to clean it if any mortar is stuck to its surface. To do this, use a brick bolster and club hammer to chip off the old mortar. Make sure you wear appropriate safety goggles and work gloves to protect yourself from flying debris when you performing this task.

Try to match the type of pointing on the existing brickwork. Refer to steps 6–9 on pages 124–5 for more details on how to create different pointing profiles.

patching rendering ⚡⚡

Many houses have a rendered finish applied to their exterior walls, either to make poor quality masonry more weatherproof or simply as a decorative feature. It is a layer of mortar applied to the surface of bricks or blocks using a two-coat system to give a final layer up to 25mm (1in) thick. The surface may be trowelled smooth or textured. It may have pebbles or other fine aggregate pressed into the surface to create the finish known as pebble-dash.

Weathering and slight movements in the masonry can eventually cause rendering to crack. This allows water to penetrate and seep down behind the rendering, and if it freezes it can cause patches to lose their grip on the masonry. These sound hollow when tapped, and may break away altogether as time goes by.

tools for the job

safety goggles

work gloves

claw hammer

chalk

brick bolster

club hammer

stiff brush

old paintbrush

bucket & mixer

hawk

pointing trowel

timber straight edge

plastering trowel

texturing tools

👍
tips of the trade

If your rendering is in generally poor condition, with a number of blown or missing areas, it will take forever to patch up and the apparently sound areas are probably on the verge of failing too. Your best bet is to hack off all the old rendering, and to call in a builder to apply a fresh coat. This job is beyond all but the most dedicated DIY enthusiast because of the large quantities of materials involved and the need for scaffolding to provide safe access for the work to be carried out.

1 Identify areas of rendering that have 'blown' – lost their bond to the masonry behind – by tapping the surface with the handle of a hammer. Mark around any areas that sound hollow with chalk. Use a ladder to reach rendering at first floor level.

2 Wearing safety goggles and gloves, use a brick bolster and club hammer to chop away the rendering until you reach a sound edge. Undercut the edges of the hole slightly to help key the new mortar patch to the old rendering, then brush all loose material out of the hole, paying particular attention to the bottom edge where most of the debris will collect.

3 Mix some sealant solution by diluting 1 part pva (polyvinyl acetate) building adhesive with 5 parts water. Use an old paintbrush to apply this sealant onto the masonry and the cut edges of the rendering. This will help the patch of new render to bond well to the masonry, and will also help to stop it from drying too quickly and cracking as it does so.

4 Mix some mortar in the ratio 1 part cement to 4 parts sharp (concreting) sand, and add some liquid plasticizer to the mix to improve its workability. For small repairs, buy a bag of dry ready-mixed mortar for rendering and just add water. Load

some mortar onto a hawk and start trowelling it into the hole. Fill the edges first, then the centre, aiming to leave the first coat about 6mm (¼in) below the surface of the surrounding rendering. Score the surface with the corner of the trowel in a criss-cross pattern to provide a key for the second coat of render.

5 Apply the second coat so that it finishes a little proud of the surrounding rendering, then use a timber straight edge to rule off the excess mortar. Hold its edge against the wall just below the patch and move it slowly upwards, while at the same time moving it from left to right in a see-saw action. This removes excess mortar without any risk of disturbing the underlying patch.

6 Fill any hollows revealed by the ruling off with a little more mortar, then float the surface of the repair smooth with a plastering trowel, wetting it first to stop the mortar from sticking to it as you work. Make the mortar as smooth as possible. If the existing surface has a texture, imitate it on the patch using whatever tools

are appropriate – perhaps a stiff brush, a sponge, a textured paint roller or a pointing trowel. If the house is pebble-dashed, load some matching pebbles onto a hawk and push them off it and into the rendering with a plastering trowel. Use the flat of the trowel to bed them well into place.

PAINTING RENDERING

If you have bare rendering, it will be much more weather resistant – and look better – with a coat of masonry paint applied to it. Kill any green algal growth on north-facing walls with a proprietary fungicide, then apply a coat of stabilizing solution to the rendering to seal its surface and reduce its porosity. You can then apply the paint, using a brush, a long-pile roller or a spray gun. The last is the quickest method, but you need to employ special spraying equipment – the fillers used in masonry paints will clog an airless spray gun. You will also have to spend time masking doors, windows, eaves, woodwork and downpipes, but the effort will be worth it, especially if you have pebble-dash or heavily textured rendering, which is time-consuming to paint by brush or roller because of the need to work the paint into all the crevices. If you do not have time to paint the whole house in one go, try to do it wall by wall rather than stop in the middle of a wall, since the joint may show up in the finished surface.

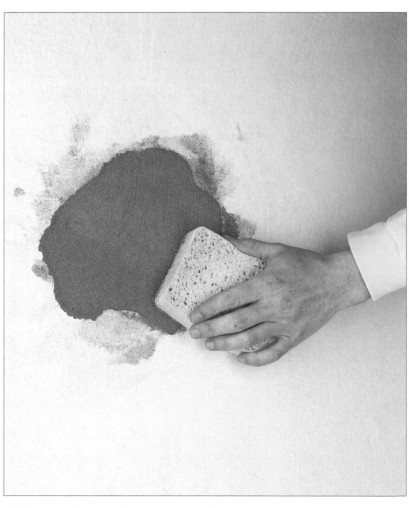

Finish the patched area of render to match the existing surface. This could mean using a sponge to texture the surface, as in this example, or perhaps applying pebble-dash.

patching concrete ↗

Concrete is the most widespread – and economical – material for laying drives, paths, patios and other surfaces around the house and garden. Although it produces a highly durable surface, it can crack if overloaded by heavy vehicles, or if the ground subsides or swells beneath it. These cracks let in water that then freezes, widening the crack and opening up further cracks that can eventually lead to the complete break-up of the surface.

If you have a concrete surface that is looking a little the worse for wear, inspect it more closely to get an idea of the true extent of the damage. Cracks, potholes and broken edges are all relatively straightforward to repair using the techniques outlined in the step-by-step sequence below, but if the surface has cracked and partly subsided, you will probably be better off breaking up the sunken section and laying new concrete in its place (see the box on the opposite page for advice on how to deal with subsidence).

tools for the job

safety goggles

work gloves

brick bolster

cold chisel

club hammer

pickaxe

spade

bucket & mixer

pointing trowel

tamping beam (for larger repairs)

plastering trowel

stiff broom

rake

1 Break away all loose material from the site of the damage, using a cold chisel, a brick bolster or even a pickaxe if necessary. Always wear protective safety goggles and thick work gloves to do this kind of work. Pick the debris out of the hole or crack piece by piece. Do not disturb any hardcore that was

beneath the concrete. If there is none and the slab was laid directly on the earth, excavate to a depth of about 100mm (4in).

2 If you had to dig out the subsoil in step 1, pack some hardcore (pieces of broken brick, flints from the garden or similar solid material) into the hole and tamp it down firmly with the handle of the club hammer.

3 Mix some concrete using 1 part cement to 3.5 parts all-in aggregate (combined sharp sand and gravel). For small patching jobs, buy a dry ready-mixed bag of concrete and just add water. Shovel it into the hole, or place it alongside a crack.

4 Tamp the concrete down into the hole using a piece of timber that is long enough to span it on edge. By moving this from side to side as you work your way across the patch, you will automatically rule off excess concrete. Fill cracks using a pointing trowel, pressing the concrete well down into the crack and finishing it flush with the surrounding surface.

5 Tamping with a timber straight edge will leave the repair with a ribbed surface finish. Flatten this with a plastering trowel if you want a smooth finish, or use a stiff-bristled broom to create a more definite textured effect. Cover the repair with

polythene sheeting weighted down with bricks. Allow it to harden for 24 hours before walking on it, and for 48 hours before driving a car over it.

DEALING WITH SUBSIDENCE

If an area of concrete has cracked and subsided slightly, you cannot simply lay more concrete on top of it to bring it back level with the rest of the slab because a thin screed will always be prone to break up and delaminate. Break up the area that has subsided using a pickaxe or a hired concrete breaker, then excavate to a depth of at least 150mm (6in) and pack in a 150mm (6in) thick layer of hardcore or crushed aggregate. The latter is better than hardcore because it compacts more fully and provides a denser and more stable base for the new concrete. Peg some timber formwork around the area you are replacing, and add more aggregate to fill the edges and corners of the mould, tamping it down well. Mix some concrete using 1 part cement to 3½ parts combined aggregate, shovel it into the mould and rake it out. Over-fill the mould slightly, then rest a tamping beam across the formwork and tamp the concrete down level with it. This will also remove any high spots. Fill any hollows that appear, and tamp again. Then finish the concrete surface to match the rest of the slab, cover it with polythene sheeting and allow it to harden for 48 hours.

6 If an edge of the concrete has broken away, remove all loose material and cut the concrete back to a sound vertical edge. Cut a piece of wood or board deep enough to match the thickness of the concrete

and long enough to span the patch, and hold it in position against the edge of the concrete with a couple of bricks or with timber pegs driven into the ground.

Pack concrete into the mould, tamping it down well with a timber straight edge to ensure that the mould is completely filled and well compacted. Finish the surface of the patch to match the surrounding concrete, cover it with polythene sheeting and allow it to harden for 24 hours. Before removing the timber the next day, run the blade of a pointing trowel between the timber and the edge of the concrete.

Use a timber straight edge to tamp the patch of concrete down well in order to achieve a flush, even surface that is safe to walk on and attractive to look at.

levelling uneven paving ✎

Paving slabs and interlocking paving blocks are very popular for giving hard surfaces around the garden a decorative finish. Although level when first laid, both slabs and blocks can become uneven due to ground movement, subsidence of the base on which they were laid or overloading by heavy traffic. Once this occurs, every raised edge is a trip hazard and the whole area soon begins to look unsightly, with weeds starting to grow through cracked joints.

Uneven paving is often the result of inadequate preparation and incorrect techniques when the slabs or blocks were first laid, especially if they were placed dry on a sand bed with joints also filled with sand. Water running off the surface of paving slabs can slowly erode the sand beneath, allowing the paving to settle. Settlement is also a common problem with block paving that was laid by hand, rather than being bedded into its sand base with a plate vibrator. The solution to this problem is to lift and re-lay the affected slabs or blocks.

tools for the job

safety goggles & work gloves

pointing trowel

brick bolster & scrap wood
(for lifting slabs)

shovel

club hammer

bucket & mixer

bricklaying trowel

spirit level

long timber straight edge

chalk

rake & stiff broom

cold chisel

dealing with uneven paving slabs

1 If a paving slab has subsided, rake out any sand or mortar pointing between the affected slab and its neighbours. Alternatively, use the tip of a pointing trowel to do this.

2 Drive a brick bolster into one of the gaps, lay a piece of scrap wood on the edge of the adjacent slab to protect it, and use the wood as a fulcrum to lever the slab up. Slide the wood underneath the raised edge so that you can get a grip on the slab and lift it out. Wear a pair of thick work gloves to give your fingers some protection while lifting out the heavy slab.

3 If the slab was laid on sand, add a little fresh sand to the bed and replace the slab. Stand it on edge at one side of the hole in which it fits, lower it to the horizontal and slide it slightly so that it drops into place without disturbing the sand

bed. Tamp it down with the handle of a club hammer until it sits level with the neighbouring slabs. If it still sits a little low, repeat the process to add more sand.

If the slab was laid on mortar dabs, chip the old mortar off the back of the slab with a brick bolster and club hammer. Take care not to crack the slab – it does not matter if a little mortar stays stuck to it. Then mix a small batch of bricklaying mortar (a bag of dry ready-mixed mortar is ideal for this sort of job) and use a bricklaying trowel to place five dabs on the bed of the hole – one near each corner and one in the middle. Drop the slab back into place and tamp it down with a hammer to the required level.

4 Lay a spirit level across the relaid slab and its neighbours to check that they are correctly aligned, then fill the joints around the relaid slab with sand or dryish mortar to match the rest of the area. If you are using mortar, lay strips of cardboard alongside the joint to prevent mortar from staining the slabs while you fill the joint.

dealing with uneven paving blocks

1 Identify the extent of the subsidence by laying a long timber straight edge across the paving in different directions. Mark any blocks that have sunk below the straight edge with chalk.

2 Wearing protective gloves, prize up several blocks at the edge of the area of sunken blocks. Work your way block by block towards the area that has subsided, and lift these blocks too.

3 Shovel some fresh sand onto the affected area and rake it out evenly, then start replacing the blocks one by one. Tamp each block down firmly into the sand bed so that it is flush with the neighbouring blocks. Use the straight edge to check that you have eliminated the subsidence after replacing every four or five blocks.

4 Finish replacing the blocks, check they are level once again and complete the job by brushing fine, dry sand into the joints. This will help to lock the blocks together and prevent subsidence from recurring.

repointing crazy paving

Crazy paving is usually bedded in mortar, so it is not generally prone to subsidence. However, the pointing between the individual stones can crack and break out, leaving unsightly paving that can also be a trip hazard. The solution is to repoint the joints with fresh mortar.

If the individual pieces of stone have begun to crack and break up, lift them and replace them with new pieces of stone, trimmed to fit.

1 Wearing safety goggles and a pair of thick protective work gloves, cut out all the failed pointing with a cold chisel and club hammer. Take care not to disturb the individual stones of the paving when you do this unless you are planning to replace damaged pieces. Brush away the debris.

2 Place new pieces of stone on a mortar bed and tamp them down level with their neighbours. Point the joints with a dryish bricklaying mortar, taking care not to get any on the faces of the stones. Fill the joint almost flush with the surface, then draw the tip of a pointing trowel along each edge of the joint in turn, letting it follow the contours of the stones. The aim is to highlight the edges of the stones and give the pointing a raised V-shaped profile.

dealing with rising damp ⚲

Houses built since the late 19th century have a waterproof layer called the damp-proof course (dpc) built into their walls just above ground level to stop ground water from being absorbed into the masonry. In older houses, the dpc is either a double layer of slate or a couple of courses of dense water-resistant engineering bricks. In modern houses it is a strip of strong plastic. A similar damp-proof membrane is incorporated in the structure of solid concrete ground floors.

tackling dpc bridges

Rising damp shows up as damp patches on interior walls, usually rising to a height of about 1m (3ft) above floor level. However, finding this type of dampness does not necessarily mean that the dpc has failed. It could have been bridged in some way, allowing ground water to bypass it and rise into the wall structure. The illustration below shows some common causes of dpc bridges. The step-by-step sequence at the bottom of the page outlines how to tackle them.

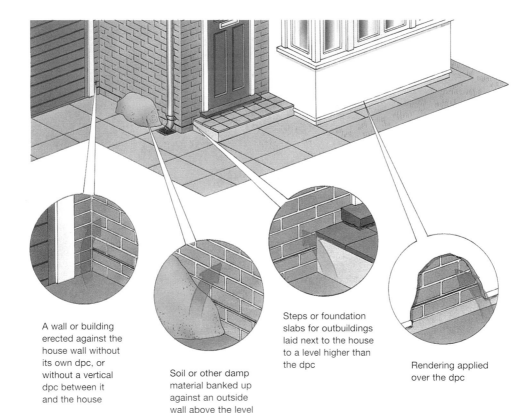

A wall or building erected against the house wall without its own dpc, or without a vertical dpc between it and the house

Soil or other damp material banked up against an outside wall above the level of the dpc

Steps or foundation slabs for outbuildings laid next to the house to a level higher than the dpc

Rendering applied over the dpc

1 Begin by clearing soil or other material that has been piled against the walls above dpc level, and make sure that airbricks have not been blocked. These are essential for providing ventilation to suspended timber floors, which can develop rot if they become damp.

2 Where walls or slabs have been built against the house and they bridge the dpc, chisel out the mortar or concrete between the two and insert a vertical dpc. Then seal the joint with non-setting mastic.

3 Treat brickwork above a path or patio that is less than 150mm (6in) below dpc level with two coats of clear silicone water-repellent sealer to stop it from absorbing water. Alternatively, lift one row of slabs next to the house and replace them with a layer of gravel.

4 Cut away rendering to just above the level of the dpc with a brick bolster and club hammer. The dpc is usually visible in the mortar joint in which it is bedded. Fix a steel and expanded metal mesh external render stop bead to the wall, level with the dpc, and render down to it to leave a neat horizontal edge.

dealing with a failed dpc

If your damp problem persists despite tackling any possible damp bridges, then it is likely that the dpc has actually failed. You can call in a damp treatment firm to install a new dpc for you, who will inject waterproof chemicals into the masonry at dpc level. However, since the work is very labour intensive and you can hire the same injection equipment the specialists use, it makes sense to do the job yourself. The tool hire companies that supply the equipment also stock the chemicals you need – you buy them on a sale-or-return basis when you hire the equipment.

Before you begin you need to find out whether you have solid or cavity walls by examining their structure – if brickwork is exposed and it consists entirely of bricks laid end to end, it is a cavity wall. If the bricks are not visible, measure the wall thickness – it will be around 240mm (9½in) if it is solid, and about 290mm (11½in) if it is a cavity wall.

tools for the job

dpc injection machine (hired)
hammer drill
long masonry drill bits
depth stop
pointing trowel

1 Hire the dpc injection machine and collect enough dpc fluid for the job. You may need up to 3 litres per square metre (5¼ pints per 10sq ft) of wall if the brickwork is very porous. Hire a professional-quality drill, too – you may well overload a DIY model. Buy masonry drill bits in a diameter to match the size of the injection nozzles and long enough to drill to a depth of 200mm (8in). Make sure the hire company shows you how to operate the machine.

2 Carry out the first stage of the injection process by drilling holes 75mm (3in) deep at about 150mm (6in) intervals along the wall at dpc level. The injecting may be done from the inside or outside. Insert the shorter nozzles supplied, secure them in place and start injecting the fluid. When complete, move the linked nozzles along to the next set of holes and repeat.

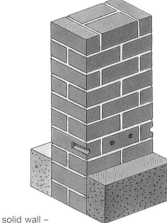

solid wall – stage 1

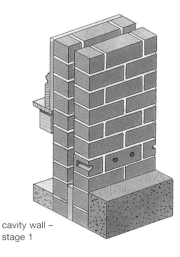

cavity wall – stage 1

3 When you have injected all the holes, drill through the same holes to a depth of 150mm (6in) in solid walls and 200mm (8in) in cavity ones. Repeat the injection process using the longer nozzles supplied with the machine. When injection is complete, fill the holes with mortar.

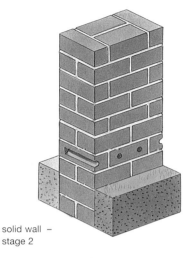

solid wall – stage 2

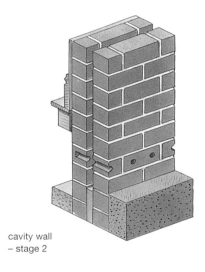

cavity wall – stage 2

tips of the trade

- **Internal walls** – If solid internal walls also show signs of rising damp, lift floorboards alongside them and inject fluid using holes about 50mm (2in) deep in a one-brick thick wall.

- **Damaged plaster** – If plaster has been damaged by rising damp, hack it off and replace it with a waterproofed cement rendering. Cover this with a coat of finish plaster when the wall has dried out.

glossary

Aggregate – broken stone, gravel or similar inert material that forms the largest part of compounds such as concrete and mortar. The finest aggregates are better known as sand. Aggregate is also sometimes referred to as ballast.

Airbrick – a perforated brick built into a wall to ventilate a room or an under-floor void.

Angle bead – a moulding made from galvanized steel and expanded metal mesh that is used to provide reinforcement for plaster on external corners of walls.

Batten – strips of wood, 50 x 25mm (2 x 1in) or less in cross-section, used to form bays on walls when plastering, and for many other purposes.

Block – a masonry unit, usually larger than a brick, used to build house and garden walls. Lightweight insulating blocks are used for cavity walls, while denser reconstituted stone blocks and pierced screen walling blocks are employed in gardens.

Block paver – brick-sized paving unit, made in a variety of colours and designs, that is laid on sand in an interlocking pattern to form driveway, path and patio surfaces.

Bond – the way in which bricks are arranged in a wall. Different bonds are used for different types of construction.

Brick – a masonry unit made from burnt clay and other materials that is used for building house and garden walls and other structures. A standard brick measures 215 x 102 x 65mm (8½ x 4 x 2½in).

Caulk – see mastic.

Cavity wall – house wall consisting of two layers of masonry held together with metal or plastic wall ties, and with a gap (the cavity) – commonly 50mm (2in) wide – between them. The inner wall is usually built with blocks, and the cavity is partially or fully filled with insulating materials.

Ceiling centre – ornamental plaster or foamed plastic moulding fixed in the centre of a ceiling for decoration.

Cement – a binder in powder form that bonds sand or aggregate together to form, respectively, mortar and concrete.

Cobbles – rounded pebbles, up to 75mm (3in) in diameter, laid loose or in a mortar bed as a garden feature area.

Concrete – a mixture of cement, sand, aggregate and water that sets to a hard, stone-like mass and is used for wall foundations and as cast slabs for patios, driveways and the bases on which outbuildings stand.

Coping – a brick, stone or concrete strip, usually overhanging, set on top of a wall to protect it from the weather.

Coving – a plain moulding, generally concave in cross-section, fitted in the angle between a wall and ceiling for decorative purposes. The term cornice refers to more ornate types of coving.

Crazy paving – a paved surface created by bedding irregular-shaped pieces of stone or paving slab in mortar and then pointing the gaps between the pieces.

Damp-proof course (dpc) – a continuous layer of impervious material (formerly slate, now usually plastic) built into a house wall just above ground level to stop ground water from soaking into the wall and causing rising damp. Concrete ground floors incorporate a damp-proof membrane (dpm).

Dry lining – a wall lining formed from tapered-edge sheets of plasterboard fixed to a framework of timber battens on the wall. It has its joints taped and filled but does not require plastering.

Efflorescence – powdery white salts left on a wall surface as it dries out after construction or plastering. They should be brushed, not washed, off.

English bond – an arrangement of bricks in a wall with alternate courses laid as headers and stretchers. The wall is one brick (215mm/8½in) thick.

Expanded metal mesh – perforated metal sheet or strip used to support plaster when patching holes in partition walls and ceilings.

Flemish bond – an arrangement of bricks in a wall where each course consists of a header followed by a pair of stretchers. The wall is one brick (215mm/8½in) thick.

Formwork – timber boards fixed to pegs in the ground that form a mould for a cast concrete ground slab. The formwork is removed when the concrete has set.

Foundation – a strip of concrete cast in a trench to support a wall or other masonry feature. It is sometimes referred to as a footing.

Gauge rod – a bricklaying aid made by marking brick and joint heights on a timber batten. It is used to check that courses in a wall are being built evenly.

Gravel – washed river stone typically sieved to a maximum stone diameter of 20mm ($^3/_4$in). It is laid loose to form a driveway or garden feature.

Hardcore – broken brick, concrete and other masonry that is laid on and rammed into subsoil to form a stable base for concrete.

Hawk – a metal or plywood square with a handle underneath that is used to carry small amounts of mortar or plaster to the work area.

Header – a brick laid in a wall with only its ends visible.

Joint tape – paper or mesh tape used to reinforce the joints between sheets of plasterboard on walls and ceilings.

Joist – a wooden or steel beam supporting a floor and, in upstairs rooms, the ceiling below.

Joist detector – electronic metal detector that finds the line of nails fixing floorboards or ceiling boards in place, and hence the joist positions.

Keystone – the central brick or stone at the top of an arch.

Lath and plaster – a lining for ceilings and stud partition walls in older houses, consisting of plaster applied to closely spaced wooden strips (the laths) that are nailed to the ceiling joists or wall studs.

Lintel – a steel, wood or concrete beam spanning the opening of a door or window.

Masonry nail – hardened steel nail that can be driven into masonry with a hammer.

Mastic – non-setting filler used to seal joints between building components,

such as between a window frame and the surrounding masonry. It is also known as caulk.

Mortar – a mixture of cement, sand and sometimes other additives used for bricklaying and rendering.

Needle – a short horizontal timber or steel beam inserted through a wall and supported by adjustable steel props in order to carry the wall's weight while part of the wall is being removed.

Nogging – short horizontal timbers fixed between wall studs or ceiling joists to stiffen the structure.

Padstone – a masonry unit used to support the end of a beam or lintel in a wall.

Paving slab – a masonry unit, usually square or rectangular and 35mm–50mm ($1^1/_2$–2in) thick, that is laid on sand or mortar to form a patio, path or driveway.

Penetrating damp – moisture entering a building through some defect in its structure and waterproofing.

Pier – a buttress projecting from one or both sides of a wall to increase the wall's stability. Piers may be built at the free-standing ends of the wall and at regular intervals along its length.

Plaster – a powder mixed with water to form a plastic material that is applied to wall and ceiling surfaces and hardens to form a smooth surface suitable for decorating.

Plasterboard – a sheet material formed by sandwiching a plaster core between sheets of strong paper. It is used for lining ceilings and stud partition walls.

Pointing – the mortar filling the spaces between the bricks in a wall. It is formed into different edge profiles using a variety of pointing tools.

Props – adjustable telescopic steel tubes that are used to support needles or the floor above when removing all or part of a wall.

PVA (polyvinyl acetate) building adhesive – white liquid used as an adhesive and sealer in building work.

Queen closer – a brick cut in half along its length and used to maintain the bond pattern in English and Flemish bond.

Rendering – a coat of mortar applied to an exterior wall surface and given a smooth finish. Pebble-dashing is formed by pressing small stones into the rendering while it is still wet.

Riser – the upright face of a step.

Rising damp – moisture entering the building from the ground due to the failure of the damp-proof course in a wall or the damp-proof membrane in a concrete floor.

RSJ (rolled steel joist) – an I-section steel beam used to support a floor or the upper part of a wall when existing supporting masonry is removed.

Sand – fine aggregate mixed with cement to form mortar. Coarse (sharp) sand is used for concreting, while finer (soft) sand is preferred for bricklaying and rendering.

Screen wall block – a square building block pierced with cut-outs and used to form decorative screen walls in gardens where complete privacy is not required.

Self-smoothing compound – a powder mixed with water and other additives to form a liquid coating. It is poured onto uneven concrete floors and left to find its own level before hardening to a smooth surface.

Springing point – the point at which the curve of an arch begins.

Stretcher – a brick laid in a wall with its side faces visible. In stretcher bond brickwork, bricks are laid end to end, with each brick centred over the joint between the bricks in the course below.

Stud partition – a timber-framed wall consisting of vertical members (the studs) nailed between horizontal head and sole plates, and clad with plasterboard on both sides.

Tread – the level part of a step.

Trowel – any one of several tools used for bricklaying and plastering. A bricklayer uses a large bricklaying trowel and a smaller pointing trowel, while a plasterer uses a rectangular steel trowel, a gauging trowel and internal and external corner trowels.

Undercoat – the first coat of plaster applied to a wall or ceiling, also known as the base coat. It is followed by a thinner top coat of finish plaster.

Wall extension profile – system of wall-mounted track and interlocking metal ties that is used to bond new masonry at right angles to an existing wall.

Wall plug – plastic insert fitted into a hole in a wall to take a screw and make a secure fixing.

Weedproof membrane – sheet material laid over subsoil beneath gravel or decking to prevent weed growth while allowing rainwater to drain through it.

index

useful contacts

suppliers

The author, photographer and publisher would like to thank the following companies:

Aristocast Originals Limited
14A Ongreave House
Dore House Industrial Estate
Handsworth
Sheffield
S13 9NP
Tel. 0114 2690900
for plaster coving

Bradfords Building Supplies Limited
96 Hendford Hill
Yeovil
Somerset
BA20 2QT
Tel. 01935 845245
www.bradfords.co.uk
for building materials

Hewden Plant Hire
Tel. 0161 8488621

Lafarge Plasterboard Limited
Marsh Lane
Easton-in-Gordano
Bristol
BS20 0NF
Tel. 01275 377773
www.lafarge-plasterboard.co.uk
plaster products

Screwfix Direct
Tel. 0500 414141
www.screwfix.com
for tools & fixings

Somerset Fireplaces
Unit 4
Horsepond Lane
Friarn Street
Bridgwater
Somerset
Tel. 01278 423795

For information on your nearest DIY superstore, contact the following:

B&Q DIY Supercentres
Tel. 0845 3002902

Dulux Decorator Centres
Tel. 0161 9683000

Focus Do It All
Tel. 0800 436436

Great Mills
Tel. 01761 416034

Homebase Ltd
Tel. 020 87847200

associations

National Home Improvement Council
Tel. 020 78288230

Brick Development Association
Tel. 01344 885651

British Cement Association
Tel. 01344 762676

British Wood Preserving and Damp-proofing Association
Tel. 020 85192588

Builder's Merchants' Federation
Tel. 020 74391753
advice on building materials and list of suppliers

Building Research Establishment
Advisory Service
Tel. 01923 664000

Federation of Master Builders
Tel. 020 72427583

Health and Safety Executive
Tel. 0541 545500

Hire Association Europe
Tel. 0121 3777707
equipment hire

Institution of Structural Engineers
Tel. 020 72354535

The Ready-Mixed Concrete Bureau
Tel. 01494 791050

Royal Institute of British Architects
Tel. 020 75805533

the author

A long-standing and highly respected writer in the field, Mike Lawrence is the author of numerous DIY project books and manuals. He is also currently the technical consultant on the highly successful BBC1 television programme 'Changing Rooms'.

acknowledgements

The photography team would like to thank the following individuals for supplying props, advice and general help throughout the production of this book – Tim Ridley, Marina Sala, Jakki Dearden, Adele Parham, Emmanuelle Baudouin, David Bevan at Aristocast, Trevor Culpin at Screwfix, David Hayward, Mark Eminson at Bradfords, John and Margaret Dearden, David House at Hewden Hire in Bruton, Michael and Sue Read, Martin and Mandy Tilly, Michael and Judith Levett, June Parham, Simon and Sandra Levett, and Johnny Koolang.
The Publisher would like to give special thanks to Axminster Power Tools and Richard Burbidge.

First published in 2002 by Murdoch Books UK Ltd
Copyright© 2002 Murdoch Books UK Ltd

ISBN 1 85391 974 8
A catalogue record for this book is available from the British Library.

All photography by Tim Ridley & Step Editions, copyright Murdoch Books UK Ltd except:
pp6, 10–11, 24, 25 (middle), 56–7, 138 Howard Rice; pp7, 20–1, 22, 23 (middle), 38–9, 92–3, 108–9, 137 Ray Main;
pp8, 25, 76–7, 85 Juliette Wade; p23 (top) Richard Burbidge; p25 Marcus Harpur; p26 (angle grinder) Axminster Power Tools.

Commissioning Editor: **Iain MacGregor**
Series Editor: **Alastair Laing**
Project Editor: **Michelle Pickering**
Designer: **Tim Brown**
Design Concept: **Laura Cullen**

Managing Editor: **Anna Osborn**
Design Manager: **Helen Taylor**
Photo Librarian: **Bobbie Leah**
Photography: **Tim Ridley & Step Editions**
Illustrations: **Mike Badrocke**

CEO: **Robert Oerton**
Publisher: **Catie Ziller**
Production Manager: **Lucy Byrne**
International Sales Director: **Kevin Lagden**

Colour separation by Colourscan, Singapore
Printed in Singapore by Tien Wah Press

Murdoch Books UK Ltd
Ferry House, 51–57 Lacy Road, Putney
London, SW15 1PR, UK
Tel: +44 (0)20 8355 1480
Fax: +44 (0)20 8355 1499
Murdoch Books UK Ltd is a subsidiary of
Murdoch Magazines Pty Ltd.

UK Distribution
Macmillan Distribution Ltd
Houndsmills, Brunell Road
Basingstoke, Hampshire, RG1 6XS, UK
Tel: +44 (0) 1256 302 707
Fax: +44 (0) 1256 351 437
http://www.macmillan-mdl.co.uk

Murdoch Books®
GPO Box 1203
Sydney, NSW 1045, Australia
Tel: +61 (0)2 8220 2000
Fax: +61 (0)2 8220 2020
Murdoch Books® is a trademark of
Murdoch Magazines Pty Ltd.